AF839697

Influence of Distribution System Infrastructure on Bacterial Regrowth

The mission of the Awwa Research Foundation (AwwaRF) is to advance the science of water to improve the quality of life. Funded primarily through annual subscription payments from over 1,000 utilities, consulting firms, and manufacturers in North America and abroad, AwwaRF sponsors research on all aspects of drinking water, including supply and resources, treatment, monitoring and analysis, distribution, management, and health effects.

From its headquarters in Denver, Colorado, the AwwaRF staff directs and supports the efforts of over 500 volunteers, who are the heart of the research program. These volunteers, serving on various boards and committees, use their expertise to select and monitor research studies to benefit the entire drinking water community.

Research findings are disseminated through a number of technology transfer activities, including research reports, conferences, videotape summaries, and periodicals.

Influence of Distribution System Infrastructure on Bacterial Regrowth

Prepared by:
Jonathan Clement, **Catherine Spencer**, and **A.J. Capuzzi**
Black & Veatch
Anne Camper and **Kristin Van Andel**
Montana State University
and
Anne Sandvig
Sandvig Consulting

Sponsored by:
Awwa Research Foundation
6666 West Quincy Avenue
Denver, CO 80235-3098

Published by the
Awwa Research Foundation
Advancing the Science of Water®

IWA Publishing

DISCLAIMER

This study was funded by the Awwa Research Foundation (AwwaRF). AwwaRF assumes no responsibility for the content of the research study reported in this publication or for the opinions or statements of fact expressed in the report. The mention of trade names for commercial products does not represent or imply the approval or endorsement of AwwaRF. This report is presented solely for informational purposes.

Published by IWA Publishing, Alliance House, 12 Caxton Street, London SW1H 0QS, UK

Telephone: +44 (0) 20 7654 5500; Fax: +44 (0) 20 7654 5555; Email: publications@iwap.co.uk;
Web: **www.iwapublishing.com**

AwwaRF report number 90950F. Originally published by AwwaRF for its subscribers in 2003.
IWA Publishing version published 2004.

ISBN 1 84339 858 3

CONTENTS

LIST OF TABLES ix

LIST OF FIGURES xi

FOREWORD xiii

ACKNOWLEDGMENTS xv

EXECUTIVE SUMMARY xvii

CHAPTER 1: INTRODUCTION 1
- Background and Project Objectives 1
- Study Overview and Report Organization 1

CHAPTER 2: LITERATURE REVIEW AND BACKGROUND INFORMATION 3
- Background 3
- Overview of Factors Impacting Regrowth 4
  - Distribution System Materials 4
  - Nutrient Levels 5
  - Disinfection Residual Levels 7
  - Temperature and Watershed Dynamics 8
  - Hydraulic Conditions 8
  - Corrosion 9
- Summary 13

CHAPTER 3: CASE STUDIES OF REGROWTH EVENTS RELATED TO MATERIALS USE 15
- Introduction 15
- Preliminary Survey 15
- In-Depth Case-Studies 17
  - Approach 17
  - Findings 18
- Summary 21

CHAPTER 4: LABORATORY STUDIES OF MATERIALS AND REGROWTH 25
- Introduction 25
- Methods and Materials 25
  - Experimental Setup 25
  - Chemical Solutions Preparations 27
  - Biological Analyses 28
  - Chemical Analyses 29
  - Bozeman Water Quality Data 30
  - Statistical Analyses 30

Results 31
Total Organic Carbon Analysis 31
Influent/Effluent Heterotrophic Plate Counts 32
Influent/Effluent Total Direct Counts 36
Biofilm Heterotrophic Plate Counts 39
Discussion 42
Biofilm HPC Data 45
Comparisons of HPC Biofilm and HPC Effluent Data 46
Summary 47
Conclusions 48

CHAPTER 5: FIELD STUDIES 55
Introduction 55
Methods and Materials 55
Experimental Setup at each Utility 55
Chemical and Biological Analyses 57
Statistical Analyses 57
Results 58
Water Quality and Biofilm Heterotrophic Plate Counts for Each Utility Study ... 58
Comparisons of Biofilm Growth 62
Discussion and Summary 63

CHAPTER 6: CONTROL STRATEGIES FOR DISTRIBUTION SYSTEM REGROWTH PROBLEMS 77
Introduction 77
Study Conclusions 77
Guidance 78
Using Annular Reactors for Site Specific Testing 78
Summary 79

APPENDIX A: PRELIMINARY AND IN-DEPTH UTILITY SURVEYS 81

APPENDIX B: CASE STUDIES 87
Richland, WA 88
Revere, MA 89
Greenville, SC 91
Columbus, OH 94
Scottsdale, AZ 95
Water District No. 1 of Johnson County, Kansas City, KS 97
Town of Gilbert, AZ 99
City of Kennewick, WA 100
Orange Water and Sewer Authority, Carrboro, NC 101
Boston Water and Sewer, Boston, MA 103

APPENDIX C: RECOMMENDED PROTOCOL FOR ANNULAR REACTOR STUDIES ... 105

APPENDIX D: PIPE EXCAVATION RESULTS AND PROTOCOL.................................... 111

REFERENCES.. 115

ABBREVIATIONS... 121

LIST OF TABLES

2.1 Percentages of Cast iron, Ductile Iron, and Steel Pipe in Total Miles of Pipe From WIDB and RAS 4

3.1 Initial Survey Respondents 16

3.2 In-depth Case Study Utilities 17

3.3 Relative Numerical Scoring Factors 19

3.4 Infrastructure Evaluation 19

3.5 Operations and Maintenance Evaluation 20

3.6 Water Quality and Treatment Evaluation 21

3.7 Summary of Scores 22

4.1 Water Quality Conditions for the Four Phases of the Laboratory Experiments 26

4.2 Bozeman Water Quality Data for the Year 2000 31

4.3 Influent and Effluent DOC Levels for All Reactors in All Phases, ug/L C 33

4.4 Average Effluent DOC Minus the Average Influent DOC for All Reactors in All Phases, ug/L C 33

4.5 Statistical Comparisons of HPCs in Effluents from Reactors 35

4.6 Statistical Comparisons of Effluent HPCs Between Phases 36

4.7 Statistical Comparisons of TDCs in Effluents from Reactors for Phase 2 through 4 38

4.8 Statistical Comparisons of TDCs in Effluents Between Phases 40

4.9 Statistical Comparisons of Biofilm HPCs for Phase 1 through 4 41

4.10 Statistical Comparisons of Biofilm HPCs Between Phases 43

5.1 Cambridge Water Quality Data for Influent to Annular Reactors 58

5.2 Statistical Comparisons of Mean Log HPC Levels for Cambridge Water Department 59

5.3 Cincinnati Water Quality Data for Influent to Annular Reactor 59

5.4 Statistical Comparisons of Mean Log HPC Levels for Cincinnati Water Works 60
5.5 Greenville Water Quality Data for Influent to Annular Reactors 61
5.6 Statistical Comparisons of Mean Log HPC Levels for Greenville Water System 61
5.7 Tampa Bay Water Quality Data for Influent to Annular Reactors 62
5.8 Statistical Comparisons of Mean Log HPC Levels for Tampa Bay Water 62
5.9 Statistical Comparison of Mean Log HPC Levels on Ductile Iron Coupons 64
5.10 Statistical Comparison of Mean Log HPC Levels on Cement Coupons 64
5.11 Statistical Comparison of Mean Log HPC Levels on Epoxy Coupons 64

LIST OF FIGURES

3.1 Pipe age distribution for survey utilities ... 23

3.2 Percent unlined cast iron pipe ... 24

4.1 HPC influent and effluent densities in phase 1 ... 49

4.2 HPC influent and effluent densities in phase 2 ... 49

4.3 HPC influent and effluent densities in phase 3 ... 50

4.4 HPC influent and effluent densities in phase 4 ... 50

4.5 Total direct counts influent and effluent densities in phase 2 ... 51

4.6 Total direct counts influent and effluent densities in phase 3 ... 51

4.7 Total direct counts influent and effluent densities in phase 4 ... 52

4.8 HPC biofilm densities in phase 1 ... 52

4.9 HPC biofilm densities in phase 2 ... 53

4.10 HPC biofilm densities in phase 3 ... 53

4.11 HPC biofilm densities in phase 4 ... 54

5.1 Annular reactor set-up at Cambridge Water Department ... 66

5.2 Annular reactor set-up at Cincinnati Water Works ... 66

5.3 Inserting a new epoxy coupon at the Greenville Water System ... 67

5.4 Tampa Bay annular reactor set-up at Pinellas County Water Treatment Plant ... 67

5.5 Cambridge Water Department influent water quality to annular reactors ... 68

5.6 HPC biofilm densities from Cambridge, MA ... 68

5.7 HPC biofilm densities from Cambridge, MA – March 20, 2000 through March 9, 2001 69

5.8 HPC biofilm densities from Cambridge, MA – March 9, 2001 through June 25, 2001 ... 69

5.9 Cincinnati Water Works influent water quality to annular reactors ... 70

5.10 Cincinnati Water Works HPC levels in influent to annular reactors 70

5.11 Attached HPC levels from annular reactor coupons - Cincinnati Water Works 71

5.12 HPC biofilm densities from Cincinnati Water Works 71

5.13 Greenville Water System influent water quality to annular reactors 72

5.14 Greenville Water System HPC levels in influent to annular reactors 72

5.15 Attached HPC levels from annular reactor coupons - Greenville Water System 73

5.16 HPC biofilm densities from Greenville Water System before DAF 73

5.17 HPC biofilm densities from Greenville Water System after DAF 74

5.18 Tampa Bay influent water quality to annular reactors 74

5.19 Attached HPC levels from annular reactor coupons - Tampa Bay 75

5.20 HPC biofilm densities from Tampa Bay 75

5.21 HPC biofilm densities from all utilities 76

FOREWORD

The Awwa Research Foundation is a nonprofit corporation that is dedicated to the implementation of a research effort to help utilities respond to regulatory requirements and traditional high-priority concerns of the industry. The research agenda is developed through a process of grass-roots consultation with subscribers, members, and working professionals. Under the umbrella of a Five-Year Plan, the Research Advisory Council prioritized the suggested projects based upon current and future needs, applicability, and past work; the recommendations are forwarded to the Board of Trustees for final selection.

This publication is a result of one of those sponsored studies, and it is hoped that its findings will be applied in communities throughout the world. The following report serves not only as a means of communicating the results of the water industry's centralized research program, but also as a tool to enlist the further support of the nonmember utilities and individuals.

Projects are managed closely from their inception to the final report by the foundation's staff and large cadre of volunteers who willingly contribute their time and expertise. The foundation serves as planning and management function and awards contracts to other institutions such as water utilities, universities, and engineering firms. The funding for this research effort comes primarily from the Subscription Program, through which water utilities subscribe to the research program and make an annual payment proportionate to the volume of water they deliver and consultants subscribe based on their annual billings. The program offers a cost-effective and fair method for funding research in the public interest.

A broad spectrum of water supply issues is addressed by the foundation's research agenda: resources, treatment and operations, distribution and storage, water quality and analysis, toxicology, economics, and management. The ultimate purpose of the coordinated effort is to assist water suppliers to provide the highest possible quality of water economically and reliably. The true benefits are realized when the results are implemented at the utility level. The foundation's trustees are pleased to offer this publication as a contribution toward that end.

This report presents research conducted to evaluate the inter-relationships between piping material, organic carbon levels, and disinfectants with respect to biofilm growth in water distribution systems. Laboratory and field studies were completed using annular reactors to evaluate biofilm accumulation on various piping surfaces. Survey data on regrowth events from several utilities were collected and a discussion of utility alternatives for reducing biofilms in the distribution system is presented.

Edmund G. Archuleta, P.E.
Chair, Board of Trustees
Awwa Research Foundation

James F. Manwaring, P.E.
Executive Director
Awwa Research Foundation

ACKNOWLEDGMENTS

The authors would like to thank the Awwa Research Foundation, without whose support this project would not have been possible. A special acknowledgement is given to Project Manager Stephanie Morales and the Project Advisory Committee, comprised of Connie Schreppel, Upper Mohawk Valley Regional Water Board, Utica, NY; Edward Bouwer, Johns Hopkins University, Baltimore, MD; and Darrell Smith, Regional Water Authority, New Haven, CT. They provided useful guidance and input throughout the project, and we appreciate their dedication and interest in our work.

We would like to thank the following utilities for conducting the annular reactor field studies and providing critical data for this project:

- **Cambridge Water Department:**
  Timothy MacDonald
  Edward Dowling
  Krystyna McInally
- **Cincinnati Water Works:**
  Faye Cossins
  David Hartman
- **Greenville Water System**
  Cam Ferguson
  Marion Freeman
- **Tampa Bay Water:**
  John Troutt
  Marsha Pryor, (Pinellas County Utilities)

The utilities listed below also provided detailed case study information for their systems:

- Richland, WA
- Revere, MA
- Greenville, SC
- Columbus, OH
- Scottsdale, AZ
- Town of Gilbert, AZ
- City of Kennewick, WA
- Orange Water and Sewer Authority, Carrboro, NC
- Boston Water and Sewer Commission
- Water District #1 of Johnson County, Kansas City, KS

In addition, the following staff at Montana State University deserve recognition: Judel Buls, Phil Butterfield, Lu Goodrum and John Neuman for their help with methods and materials and Marty Hamilton for assistance in the statistical analyses of the data.

Finally, the authors would like to thank the following people for their assistance throughout the project and/or their participation in the final workshop:

Alan Hess, BV
Andrea Hargett BV
Amy Kramer, BV
Betsy Reilley-Matthews, MWRA
Windsor Sung, MWRA
John Sullivan, BWSC
Ron Hunsinger, East Bay MUD
Bryan Warwood, Biosurface Technologies
Joe Harrington, Cambridge Water Board

EXECUTIVE SUMMARY

RESEARCH OBJECTIVES AND APPROACH

The primary objectives of this study were to evaluate the influence of pipe materials/linings on bacterial regrowth in the distribution system, and to examine the inter-relationships between piping material, organic carbon levels, and disinfectants with respect to biofilm growth. This was accomplished by:

- conducting a literature review,
- gathering survey data from utilities to obtain real-world information on utility regrowth events,
- conducting controlled laboratory experiments using annular reactors to evaluate the relative importance of biodegradable organic matter and disinfectant as they relate to biofilm accumulation on various piping surfaces, and
- completing field study evaluations of biofilm growth on various piping surfaces, also using annular reactors.

The survey data was collected through a two-step process. An preliminary survey was sent to 185 utilities throughout the U.S., and from this initial information, a select number of utilities were chosen for more detailed, in-depth case study write-ups. Materials of use, water quality, and operational procedures employed by the utility were documented and evaluated.

The pipe materials investigated in the laboratory and field study portions of the project included ductile iron pipe, cement lining, epoxy and PVC. These materials were chosen because of their prevalence of use in the distribution system and potential link to regrowth events (iron, cement lining) and/or their use as an alternate material in relining or replacement efforts (PVC and epoxy). The testing of these materials was accomplished using a rotating annular reactor that contains a large group of material based coupons. The amount of bacterial regrowth on the coupons was measured by quantifying the number of HPCs over the surface area.

The objectives of the laboratory experiments were to evaluate the relative importance of two key water quality parameters; 1) the amount of biodegradable organic matter, and 2) the type of disinfectant used, as they relate to effluent counts from the reactors and biofilm accumulation on unlined iron, PVC, epoxy, and cement surfaces. The laboratory experiments utilized four pairs of annular reactors. Each reactor within a pair contained the same coupon material (ductile iron, PVC, epoxy, or cement). The experiments were conducted in four phases with varying controlled conditions. During the first phase of the laboratory experiments all reactors were treated identically. The reactors were fed biologically treated tap water and amended with nitrogen and phosphorus to maintain a carbon-limited growth condition. In the second phase, one reactor within each pair received free chlorine while the other received monochloramine to maintain a residual of 0.2 mg/L measured as free and total chlorine, respectively; all other conditions remained the same as in the first phase. In the third phase, all of the reactors were supplemented with 0.5 mg/L total carbon derived from humic substances. All other conditions remained the same as in the second phase. In the fourth phase, conditions were the same as the third phase except that the supplemented carbon level was raised to 2 mg/L total carbon.

The field study component of the project utilized annular reactors at four different utility locations; Cambridge, MA; Cincinnati, OH; Greenville, S.C; and Tampa Bay, FL. The objective of these studies was to determine the biofilm growth on coupons made of various materials in order to evaluate how water quality and pipe material interact to foster biofilm growth at each specific utility location. The materials evaluated in the field studies were ductile iron, cement, and epoxy.

FINDINGS

Analysis of the in depth survey and case studies of regrowth events found that the most important components predisposing water systems toward coliform incidents or regrowth appeared to be less than optimal water quality conditions and infrastructure. Systems with older, unlined cast iron pipe were associated with higher incidence of coliforms.

The laboratory portion of this study yielded the following results for effluent total direct counts (TDCs), effluent heterotrophic plate counts (HPCs), and biofilm HPCs:

- Densities measured using the TDC method were 10 to 100 times higher than those measured with the HPC method. This phenomenon is probably due to cell injury, variations in growth phase, and genotypic or phenotypic variations which prevented growth by the HPC method.

- Effluent TDC levels for phase 2 (disinfectants only) were ≤ phase 3 (disinfectants plus 0.5 mg/L TOC) ≤ phase 4 (disinfectants plus 2.0 mg/L TOC), and phase 2 was < phase 4 in all cases

- Effluent HPC levels for 2 (disinfectants only) were ≤ phase 3 (disinfectants plus 0.5 mg/L TOC) ≤ phase 4 (disinfectants plus 2.0 mg/L TOC) in all cases, and generally, phase 4 was equal to phase 3. Effluent HPC levels for phase 4 = phase 1 in all cases.

- Biofilm HPC levels for phase 4 = phase 1 for all materials except iron. For biofilm HPC levels for reactors containing iron, phase 4 > phase 1.

- In all cases, biofilm densities in phase 2 (disinfectants only) were the lowest or not significantly different from the lowest.

- In all cases, biofilm densities in phase 4 (disinfectants plus 2.0 mg/L TOC) were the highest or not significantly different from the highest.

- On the materials other than iron, there was no regular pattern regarding which reactors had the highest or lowest biofilm densities.

- The biofilm HPC densities in the PVC reactors were always the lowest or not significantly different from the reactors with the lowest densities.

- The biofilm HPC densities in the epoxy and cement reactors were often not significantly different from each other and had no significant pattern of ascendance.

- The iron reactors had the highest biofilm densities or were not significantly different from the highest densities in all phases. These results support previous work which has shown that the biofilm densities on iron are significantly greater than those on plastic-based materials such as PVC. This earlier work also showed that the densities on cement are intermediate to those on PVC and iron.

- Comparisons of reactors with the same materials that were treated with different disinfectants showed that in general there was no significant difference in the effluent or biofilm densities between the reactors.

- Among cases where significant differences were found between disinfectant treatments for a given material, there was no pattern to show that one disinfectant had greater efficacy than the other.

- Biofilms from the iron reactors treated with monochloramine in phase 2 (disinfectants only) were statistically significantly denser than those in phase 1 (control). Biofilms from the iron reactors treated with chlorine in phase 2 appeared to be denser than those in phase 1, but the difference was not found to be statistically significant. Biofilms in iron reactors treated with either disinfectant in phase 2 were significantly denser than biofilms on any other materials treated with the same disinfectant. In phase 1 this difference was much smaller. This evidence supports previous findings which showed that the efficacies of monochloramine and chlorine are further reduced against biofilms in the presence of iron.

- Biofilms in iron reactors which received supplementary humics at both 0.5 and 2 mg/L were significantly denser than those on any other materials under the same conditions. There was very little difference between the densities on the other materials. This evidence supports previous work which has shown that there is an increase in the bioavailability of humic substances in the presence of iron. Since the disinfectants were still being added with the supplementary carbon, the chlorine and monochloramine effects were probably also still a factor.

The field study component of this project yielded the following results:

- The iron coupons exhibited the highest attached HPC levels, or were not statistically different from the materials with the highest levels for all utility studies.

- In general, there was no statistical difference between biofilm growth on the cement and epoxy coupons.

- Iron coupons from Cambridge had significantly higher biofilm growth than the iron coupons from any of the other utility studies.

- There was no statistical difference in log attached HPC level between the three materials tested at the Miller Treatment Plant in Cincinnati or in Tampa Bay.

- The attached growth measured on the cement and epoxy coupons at Tampa Bay was higher than the growth on these materials from the other utility locations. Tampa Bay exhibited the highest influent TOC values of all the utility field studies.

- The attached growth on the cement and epoxy coupons exhibited similar ranges of values in the Greenville, Cincinnati, and Cambridge studies.

CONCLUSIONS

The laboratory experiments found no significant difference in the efficacies of chlorine and monochloramine in controlling biofilms or effluent counts at a residual of 0.2 mg/L. There was also no significant difference in the impacts of these disinfectants on either planktonic or biofilm cells as a function of material. Increases in the carbon levels fed to the reactors led to general increases in biofilm and effluent HPC concentrations, and this effect was most pronounced for biofilms in reactors containing iron. Reactors containing PVC had the lowest biofilm and effluent counts while reactors containing iron had the highest biofilm and effluent counts, regardless of the organic carbon levels. The biofilm and effluent counts in reactors containing cement and epoxy fell between the PVC and iron with no apparent order of ascendance.

The field testing results indicate that in certain conditions, iron resulted in the highest regrowth, and in other cases, the type of material had no influence on the number of bacteria present. Regrowth appeared to be very site specific depending on the nature and quantity of the organic material present and the type of piping material present. While iron coupons generally exhibited the highest biofilm levels, it should not be concluded that this would always be the case. Site specific testing should be performed to determine if the type of material in the distribution system is a contributing factor in regrowth.

GUIDANCE

The conclusions from this study have practical applications to the drinking water industry. This study has shown that higher levels of carbon supported greater biofilm growth and planktonic populations on the materials that were tested, although the effect was most pronounced on iron pipe. Utilities with significant amounts of iron pipe in their distribution systems may be faced with the greatest regrowth potential, and for utilities that also distribute water that is high in natural carbon, this problem may be compounded. Therefore, reducing the organic carbon in the finished water may be effective for any utility wishing to reduce regrowth problems in their distribution system. Other alternatives for utilities wishing to reduce biofilms in the distribution system can include any or all of the following; replacing or relining iron pipe in the system, increasing disinfectant, and implementing effective corrosion control. While evaluations of these options were outside the scope of this project, this study has demonstrated that site specific testing can help utilities determine the most cost-effective approach to solving their regrowth problems. Annular reactors, when operated correctly with analyses completed by a qualified laboratory, can be a practical tool to accomplish these evaluations.

CHAPTER 1
INTRODUCTION

BACKGROUND AND PROJECT OBJECTIVES

Solving bacterial regrowth problems has historically been linked to improving source water treatment or improving secondary disinfection in the distribution system. Most water systems have battled their bacterial regrowth problems (e.g. positive coliforms) by increasing disinfectant residual, converting from free chlorine to chloramine, or installing source water treatment processes. However, a number of distribution system related coliform regrowth occurrences were not controlled by higher disinfectant residuals, conversion to chloramines, or changes in source water treatment conditions.

There has been a growing body of data that strongly indicates that the amount of attached bacteria on the pipe can be several orders of magnitude higher on iron pipe surfaces than on other piping materials such as cement linings, PVC, or polyethylene. These data suggest that the amount of bacterial regrowth depends on the inter-relationship between the type of pipeline materials and the specific water quality in the distribution system, which is a function of the type of source water treatment.

In addition, much attention has been given to the ability of disinfectants (chlorine and chloramine) in the distribution system to reduce bacterial regrowth. Many water systems have maintained high levels of disinfectants (>2 mg/L Cl_2) without any significant reduction in bacterial regrowth. In addition to being of limited effectiveness for controlling bacterial regrowth, higher levels of disinfectants can promote disinfection by-product formation and offensive taste and odors. Therefore, other means for reducing bacteria must be examined. Many countries in Europe, particularly the Netherlands, Germany, and Switzerland have begun to reduce or eliminate chlorine by improving source water treatment and replacing iron piping materials to minimize bacterial regrowth.

The primary objective of this study was to evaluate the influence of pipe materials/linings on bacterial regrowth in the distribution system, and to examine the inter-relationships between piping material, organic carbon levels, and disinfectants with respect to biofilm growth.

STUDY OVERVIEW AND REPORT ORGANIZATION

The approach for this study included several components, and this report has been structured to present the results from each:

1. A literature review (Chapter 2)
2. Preliminary and detailed surveys to obtain real-world information on utility regrowth events. The results of these surveys were used to prepare 10 case study write-ups of specific utility experiences with coliform occurrences, and to document the role that pipeline materials, water quality, and system operations may have played in these events (Chapter 3).
3. Controlled laboratory experiments using annular reactors to evaluate the relative importance of biodegradable organic matter and disinfectant as they relate to biofilm accumulation on various piping surfaces (Chapter 4),

4. Field study evaluations of biofilm growth on various piping surfaces using annular reactors (Chapter 5.)

The results from each of these tasks were used to evaluate the interrelationships between distribution system materials and water quality conditions, and to provide guidance that can be used by water systems to help them mitigate their regrowth events. This guidance is contained in Chapter 6.

CHAPTER 2
LITERATURE REVIEW AND BACKGROUND INFORMATION

BACKGROUND

In spite of the low organic carbon concentrations, presence of disinfectants, generally low temperatures, and flow regimes, the growth and persistence of bacteria has been widely observed in drinking water distribution systems. While bacteria can grow as planktonic organisms (suspended in the bulk water), they replicate faster when grown as a biofilm on a pipe surface (Characklis and Marshall 1990). The term *regrowth* is used to describe the chronic or periodic appearance of bacteria in the distribution system, either in the bulk water or at the pipe walls (Camper 1994).

Biofilm growth may have an advantage over suspended bacterial growth because:

1) the flow in water distribution systems can transport large quantities of nutrients to attached microorganisms, even when the nutrient concentration in the water is low; and
2) microorganisms can form a capsular material of extracellular polysaccharides (EPS) that anchors them to the pipe surface. This EPS and the corresponding accumulation of corrosion products and particulates can exert a significant disinfectant demand and provide a shield that protects microorganisms from lethal levels of disinfectants (LeChevallier et al 1996, Fletcher and Marshall 1982).

The concentration of planktonic cells in distribution systems can be increased by erosion and sloughing of bacteria from these biofilms (van der Wende et al., 1988). There have been many occurrences of positive coliform bacteria episodes in water utility distribution systems where there were no obvious cross-connection problems or treatment breakthrough (Smith et al 1990). These coliform bacteria may grow in biofilms on the pipe surface, and coliforms have been enumerated from the surfaces of distribution system materials (Olson 1982; LeChevallier et al 1987; Kirmeyer et al 2001). Other microorganisms found in distribution biofilms include actinomyces, molds, fungi, nitrifying bacteria, iron oxidizing bacteria, and sulfate reducing bacteria (LeChevallier 1990; Geldreich 1986; Camper 1994).

Recent studies have shown that the most important factors impacting regrowth in water distribution systems may be the type of pipe material present in the system and the amount of organic carbon available as a nutrient for growth (Camper 1996; Abernathy and Camper, 2001). LeChevallier et al (1998) suggests that the type of pipe material may actually be a more important factor than organic carbon or the type of disinfectant applied by the water system. The chemical and physical reactions that occur at the pipe wall may impact the types of bacteria present in biofilms, their metabolic activity, and their release back into the water column.

There are many other factors that can impact regrowth in water distribution systems, including nutrient levels, presence of disinfectants, temperature, precipitation, and hydraulic regime. The following sections provide an overview of these factors and their interrelationship with piping materials and the potential for regrowth.

OVERVIEW OF FACTORS IMPACTING REGROWTH

Distribution System Materials

A variety of materials are used for the piping, storage facilities, pumps, valves, fittings and other appurtenances in water distribution systems, including lined and unlined cast iron, ductile iron, steel, galvanized iron, polyvinyl chloride (PVC), polyethylene (PE), concrete, asbestos cement, copper, and lead.

There is still a very large amount of unlined iron or steel pipe in service throughout North America. Based on data from the Water Industry Database (WIDB), over 50% of the total miles of distribution system piping in place is comprised of some type of ferrous material and approximately one quarter is unlined (Table 2.1). In addition, it is probable that the linings of many of the older cast iron and steel pipes have deteriorated and may also be susceptible to the effects of internal corrosion.

The percentages shown in Table 2.1 do not include service lines however, which quite often have stagnant water conditions, which could contribute to water quality deterioration. While the most common service line materials of use are copper and plastic, there are still a number of lead services in place as well as galvanized steel. Older galvanized steel pipes may have little or no zinc coating left.

Table 2.1
Percentages of Cast iron, Ductile Iron, and Steel Pipe in Total Miles of Pipe from WIDB and RAS [(1)]

Material	WIDB (%) (1097 utilities)
Cast iron (total)	48.0
Lined	25.8
Unlined	22.2
Steel (total)	4.0
Ductile Iron (unlined)	2.5
Total Ferrous Pipe	54.5
Total Unlined Ferrous Pipe[(2)]	24.7

Source: Data from AWWA and AwwaRF, 1992
(1) Excluding service lines
(2) The percentages of unlined steel pipe are not known; these totals do not include steel pipe.

In addition to the materials used to form the piping network, there are numerous coatings and linings that may be used in distribution systems, for both piping and storage facilities. These include cement lining and various epoxies and plastic slip liners.

Although extensive research has been conducted regarding the growth of bacteria on iron surfaces (LeChevallier et al., 1990a, Abernathy, 1998, Geesey et al, 1989, LeChevallier et al., 1993), very little research has been conducted on inert materials such as epoxy, cement, and polyvinyl chloride (PVC). Some characteristics of pipe materials that can influence the proliferation of biofilms include roughness and reactivity. In one study, similar biofilm densities were observed on PVC and polyethylene, suggesting that materials with similar porosity and roughness support similar biofilm densities. The same study showed that plastic based materials such as PVC or polyethylene support less growth than cement-based materials, while iron materials support the most growth (Niquette et al, 2000).

Several researchers have documented that unlined mild steel, followed by unlined cast iron, and unlined ductile-iron surfaces will support significantly higher biofilm densities than non-ferrous materials (LeChevallier 1997; Delanoue, et al 1997). In a comparison of biofilm densities on different materials, the number of bacteria on a cast iron pipe was highest, and PVC pipe biofilm densities were the lowest (Neden et al 1992). LeChevallier et al (1996) found that water distribution systems that contain large quantities of unlined cast iron and unlined ductile-iron pipes frequently experienced problems with coliform violations and taste and odor complaints. Camper (1996) found that mild steel supported 10 times more heterotrophs and coliforms than polycarbonate when the two systems were operated identically in a pilot study of factors impacting regrowth events. In this study, the presence of iron-based corrodible pipe materials and elevated organic carbon concentrations were the two most important parameters with respect to occurrence of regrowth events. LeChevallier et al (1998) found that biofilms grew faster and attained higher levels on iron pipes when compared to PVC pipes, and the population was more diverse. Kerr et al (1999) compared biofilm accumulations and colony morphology types on three materials; cast iron, medium density polyethylene (MDPE), and unplasticized polyvinyl chloride (uPVC) using a modified Robbins Devices (nMRD). They found that the mean doubling times and mean viable counts for heterotrophic bacterial populations were greater for cast iron when compared to MDPE and uPVC. MDPE and uPVC supported the lowest number of bacteria in a steady state biofilm and the diversity of heterotrophic bacteria was greatest on cast iron.

Nutrient Levels

Organisms must obtain the proper nutrients from their environment in order to grow and reproduce. Coliform and heterotrophic bacteria's principal nutrient sources are organic carbon, nitrogen, and phosphorus in a ratio of approximately 100:10:1 (C:N:P). Organic carbon is often considered the growth limiting nutrient.

Carbon

In drinking water supplies, organic carbon compounds consist of humic and fulvic acids, polymeric carbohydrates, proteins, and carboxylic acids (USEPA 1992). Organic compounds are present from natural processes and are frequently called natural organic matter or NOM. Concentrations of NOM in drinking water range from 1 – 20 mg/L with 2 – 4 mg/L being typical.

Biodegradable organic matter (BOM) is the term used to describe available organic carbon that can be used as a nutrient for growth and is usually a small fraction of the NOM. All measurement techniques determine only some portion of the total BOM. Two different measurement techniques have been developed to describe the organic carbon fraction in drinking water that is utilized by bacteria as a carbon source. Assimilable organic carbon (AOC) is defined as a calculated quantity of small, readily degradable of dissolved organic carbon based on the number of organisms that grow in the test water. Biodegradable dissolved organic carbon (BDOC) refers to a specific test that measures the change in total organic carbon in the water after it has supported the growth of indigenous microorganisms. The BDOC content of a given water may be greater than the AOC value due to the more diverse microbial population used in the assay. The various procedures used to obtain AOC and BDOC are summarized by Huck (1990).

Servais et al (1993) recommend a BDOC limit of 0.15 mg/L to control HPC growth in water with no disinfectant residual. van der Kooij (1992) reported a significant correlation between the regrowth of HPC bacteria and AOC levels in Dutch distribution systems, with growth limited at AOC levels less than 10 ug C/L. LeChevallier et al (1991a) correlated the occurrence of coliform bacteria in the distribution system to AOC concentrations greater than 50 ug C/L. LeChevallier et al (1987) found that AOC concentrations decreased with travel time of the water from the treatment plant, possibly due to the utilization of the carbon by biofilm organisms. The impact of removing this nutrient from the distributed water can reduce biofilm levels in the distribution system. Servais (1993) reported that systems with BDOC of 0.1 mg/L had 10 times lower biofilm levels than systems with BDOC of 0.4 – 0.7 mg/L.

Abernathy (1998) suggested that the coiled structure of humic substances may decrease the bioavailability of humic substances to suspended bacteria because the enzymatically active sites of the molecules are hidden within the coils. However, adsorption of humic substances to iron oxides has been linked to an increased ability of microorganisms to use the humic substances as a food source (Camper et al., 1999). Qi (1999) demonstrated that sorbed humic substances alone could not support biofilm accumulation, but that they can be used as a supplementary carbon source in biofilm accumulation on iron surfaces when humic substances are supplied in the bulk fluid. In low carbon environments like drinking water distribution systems, this effect may be of greater importance.

Several mechanisms have been proposed to explain the ready adsorption of natural organic matter to iron oxides. Previous studies by Gu et al. (1995) showed that different fractions of NOM are adsorbed by iron oxide with different affinities and capacities. It has been shown that on a carbon-weight basis, larger size hydrophobic NOM fractions had higher adsorption affinities and capacities than smaller size fractions. In addition, the larger sized hydrophobic fractions were preferentially adsorbed over the smaller sized hydrophilic fractions. Due to the difficulty in quantitatively identifying NOM fractions from a mixture before and after adsorption, competitive adsorption has not yet been explored (Gu et al., 1996). Another study by Gu et al. (1994) demonstrated the pH dependence of the adsorption of natural organic matter to iron oxide surfaces. At lower pHs the adsorption of the NOM was high, but it decreased rapidly as the pH increased.

Nitrogen and Phosphorus

Nitrogen in the distribution system can be in the form of organic nitrogen, ammonia, nitrate and or nitrite and all forms can be utilized by different microorganisms. Ammonia can be elevated in distribution systems due to chloramination practices, and can support autotrophic growth (nitrifiers). One study found that bacterial occurrences in the distribution system were associated with the application of nitrogen containing agricultural fertilizer that entered source water in storm water runoff (Geldreich 1986).

The occurrence of phosphorus in the environment is almost all as orthophosphate. Application of phosphate-based corrosion inhibitors may actually improve conditions in distribution systems, with respect to microbiological growth. In Seymour, Indiana, the coliform bacteria levels decreased within a few weeks after application of zinc orthophosphate (Lowther and Moser 1984). A study by Rosenzweig (1987) found that high levels of Virchem 932, a zinc orthophosphate product, inhibited growth of certain coliform species. LeChevallier et al (1991a) also found that utilities that used phosphate based inhibitors for corrosion control experienced significantly decreased bacterial growth.

Disinfection Residual Levels

Drinking water utilities in the United States are required to maintain a disinfectant residual in drinking water distribution systems. Monochloramine and chlorine are the two most commonly used disinfectants for this purpose and must be maintained at a residual of 0.2 mg/L in the distribution system. The purpose of maintaining this residual is to prevent regrowth in the distribution system and to inactivate any microorganisms that enter the distribution system as a result of contamination. However, the efficacy of this residual in preventing biofilm growth is limited as biofilms are significantly more resistant to disinfection than suspended cells of the same strain (Costerton et al., 1987).

Bacteria growing on pipe surfaces can be thousands of times more resistant to chlorine disinfection than planktonic cells (LeChevallier et al 1988a). Experiments by LeChevallier et al (1988b) have shown that attachment of bacteria to surfaces provides the greatest increase in disinfection resistance, implying that microbes entrapped in particles or adsorbed onto surfaces are shielded from disinfection and are not inactivated. Biofilm bacteria grown on GAC particles, metal coupons, and glass slides were 150 to 3000 times more resistant to hypochlorous acid (free chlorine at pH 7.0) and 2 to 100 times more resistant to monochloramine than unattached cells. The advantage of monochloramine over free chlorine in inactivating biofilms was also demonstrated in a 1990 study by LeChevallier, et al (1990a).

It has been shown that decreased chlorine efficacy against biofilms is due in part to a transport limitation resulting from a reaction-diffusion interaction in the biofilm. The microorganisms, exopolysaccharides, and other reactive biofilm constituents could consume the chlorine, protecting the biofilm beneath from exposure (Xu, 1996) (Stewart, 1998). This effect is enhanced in the presence of iron (LeChevallier et al., 1993). Monochloramine is also believed to have reduced efficacy against biofilms due to a reaction-diffusion limitation (Srinivasan et al., 1995). Like chlorine, the disinfectant efficacy of monochloramine is further reduced against biofilms in the presence of iron, though to a lesser degree than chlorine (LeChevallier et al., 1993).

Since chlorine is more widely reactive, it can be rapidly consumed by system components and materials in the water, lowering its performance as a biocide. In an annular reactor study comparing chlorine and monochloramine efficacies against biofilms, chlorine was found to be highly reactive in the uninoculated system whereas monochloramine did not react (Griebe et al., 1994).

Monochloramine is also gaining popularity for its increased efficacy against biofilms compared to that of chlorine. In one study, monochloramine was shown to be more effective against *Pseudomonas aeruginosa* biofilms than free chlorine (Griebe et al., 1994). Another study of the Greater Vancouver Water District also showed that chloramine is a more effective disinfectant for controlling biofilm growth in distribution systems. This was evidenced by decreased levels of coliform and HPC bacteria. This study also indicated that chloramine, as a secondary disinfectant, produces a more stable residual, less taste and odor, and is significantly less expensive than chlorine (Neden et al., 1992). LeChevallier et al. (1988b) found that due to its greater penetrating power, monochloramine was better able to penetrate and kill biofilms than chlorine. LeChevallier et al. (1993) also showed that monochloramine at a residual of 3.4 mg/L is significantly more effective than chlorine at a higher residual of 4.1 mg/L in the presence of iron. This was attributed to the reactivity of chlorine with corrosion products.

One downside to monochloramine as a biofilm disinfectant was shown in a study of monochloramine as a disinfectant against *Pseudomonas aeruginosa*. In this study, evidence was found that *P. aeruginosa* can adapt to prolonged or repeated treatments. It was theorized that at low doses (<0.5 mg/L) these bacteria produce monochloramine-neutralizing biomass constituents. It was also theorized that cells exposed to monochloramine acquire reduced susceptibility to disinfection. This study suggests that it is more effective to deliver monochloramine in a short concentrated dose than in a longer, less concentrated dose (Sanderson et al., 1997). This phenomenon is of importance to drinking water science as *P. aeruginosa* is common in finished waters and distribution system biofilms. Although it is not a frank pathogen, it is an opportunistic pathogen, and can cause severe respiratory and other infections in populations with weakened immune systems such as newborns, the elderly, and AIDS patients (Cohn et al., 1999).

Temperature and Watershed Dynamics

Temperature may be the most important factor controlling the rate of microbial growth because of its affect on microbial growth rate, disinfection efficiency and system hydraulics/ water velocity (due to customer demands during warmer temperature periods) (LeChevallier, 1989). Many studies have shown that bacterial numbers in the distribution system increase when water temperatures are 15°C or warmer (Fransolet et al 1985; Donlan and Pipes 1988; Smith et al 1989; LeChevallier et al 1990a; Donlan, Pipes, and Yohe 1994). In a survey of 81 water distribution systems, Smith et al (1990) observed that the highest coliform levels occurred during the summer months and that the species of coliform bacteria varied with water temperature. In warm climates or where seasonal variations in temperature are less pronounced, seasonal effects on coliforms may be less dramatic (USEPA 1992). In a study by Fransolet et al (1985) water temperature was found to influence bacterial growth rate, lag time, and cell yield, indicating that at low temperatures cells are washed out of the distribution system before significant growth is achieved.

Watershed dynamics may also be an important factor in coliform growth in the distribution system. The increase in turbidity that occurs after rainfall events may provide bacteria with particles for attachment and protect them from disinfection (USEPA 1992). Raw water organic nutrient levels have been shown to be higher after rainfall events (Lowther and Moser 1984) and LeChevallier et al (1990b) found that coliform bacteria appeared in distribution systems 7 days after rainfall events. Several New England water systems have also experienced increased coliforms after heavy rainfalls (Geldreich 1986).

Hydraulic Conditions

The hydraulic conditions in the distribution system are another factor in the dynamics of regrowth occurrence. Increased flow velocities can cause shearing of biofilms from the pipe surface, expose the biofilm to increased nutrient levels, and provide greater transport of disinfectants. Stagnant water can cause loss of disinfectant residual and accumulation of sediment, which may lead to microbial growth (USEPA 1992). Donlan and Pipes (1988) showed that water velocity had an inverse relationship with biofilm counts and dead-end lines have historically shown significant deterioration in microbial water quality (Smith, et al., 1989; Geldreich, 1986; Hanson et al., 1987; Opheim et al 1988; Kirmeyer et al 2001).

Changes in water velocity can also occur due to pipe network design and pipe size, water main breaks, or distribution system maintenance practices such as flushing (Smith et al 1989; Geldreich, 1986). Opheim et al (1988) found that bacterial levels in an experimental pipe system increased 10 fold when flows were started and stopped, and larger releases were observed when the system was subjected to physical and vibrational forces such as might occur with the water "hammer" effect upon sudden opening or closing of lines. The residence time of water in a system may be an important factor in determining where organisms are most likely to occur on pipe surfaces. Assimilable organic carbon (AOC) levels and BDOC concentrations have been shown to decrease as water moves through the distribution system (LeChevallier et al 1987; LeChevallier et al 1991a; van der Kooij 1992), although this is not universally true (Lui et al 2002).

Corrosion

Corrosion is an oxidative process that occurs at the surface of the metal. The formation of films on the insides of pipes is an extremely complex physical/chemical process that results in the dissolution of the metal and the formation of layered mineral and/or amorphous scales on the surface of the pipe. In the case of iron pipe, the pure metal is oxidized into ferrous hydroxide [$Fe(OH)_2$], which may be further oxidized to ferric hydroxide [$Fe(OH)_3$] by reaction with oxygen. The ferrous and ferric ions created from metallic corrosion are available to combine with several anions in the water (CO_3^{-2}, PO_4^{-3}, S^{-2}, OH^-, etc...) to form various soluble and precipitated corrosion by-products (tubercles). Corrosion products are of interest because systems with significant amounts of corrosion by-product mass have been found to contain substantial microbial densities (Rice, et al 1991; Herson et al 1991; van der Kooij and Oorhuizen 1997; McMath, et al 1997; Martin, et al 1982; LeChevallier, et al 1991b; LeChevallier, et al 1993; Camper 1994; Abernathy and Camper 2001).

Tubercles

Over time the oxidation reaction slows as corrosion products (tubercles) adhere to the iron surface and form a protective layer between the pure iron and the reactants in the bulk fluid (Geesey et al., 2000). As iron oxide corrosion products form on the surface of the pipe, the texture of the pipe changes from a smooth, homogeneous surface to a rough, heterogeneous surface. This rough surface provides a sheltered habitat for the growth of microorganisms into biofilms. In addition, the corrosion products react with disinfectant residuals, which prevent disinfectants from penetrating these biofilms (LeChevallier et al., 1993).

A variety of bacterial types and relatively high numbers of organisms have been found to be associated with tubercles taken from distribution system pipe walls (Allen et al 1980; Tuovinin and Hsu 1982; Martin et al 1982). In the study by Tuovinen and Hsu (1982), aerobic and anaerobic bacteria were found in tubercles collected from sections of water distribution system pipelines in Columbus, Ohio where the bulk water chlorine residuals were 0.2 to 1.0 mg/L free chlorine. Sulfate reducing bacteria comprised 80% of these samples and nitrate reducing heterotrophs were present in all samples. The tubercle mass contained 50% Fe(II) with a dense layer of magnetic iron oxide (magnetite) comprising most of the surface crust.

High levels of heterotrophic bacteria, but no coliforms, were obtained from tubercle and sediment material from pipes in the New Haven, Connecticut water system (McFeters and Camper 1988). Iron bacteria and sulfate reducing bacteria were found to be associated with these iron tubercles. Coliform bacteria were found in iron tubercles from a 28 year old cement-lined cast-iron main where parts of the lining had broken away (LeChevallier et al 1987). The bacteria isolated from the tubercles had the same biochemical profile as organisms in the water column. High levels of coliforms were also detected in iron tubercles taken from the distribution system in this study, and all materials (cast iron, cement lined, ductile and asbestos cement) had high heterotrophic plate counts. This research showed that coliforms can be associated with tubercles within the distribution system and that organisms growing on these tubercles can be shielded from disinfection. Any physical or chemical attack on these tubercles could result in release of coliform and HPC bacteria into the water column.

In another study, scale and biofilm evaluations were completed on unlined and lined cast iron, galvanized iron, and lead pipe specimens. Coliforms were found in nine of the 18 pipe deposits analyzed, including three of the seven lead pipes as well as from the cement lining of lined cast iron pipe. No coliforms were found on the galvanized iron pipes. Sulfate reducing bacteria were present in all but one sample from ferrous metal pipes and three samples from the lead pipes. There was a considerable range in the numbers of heterotrophs enumerated in the deposits but no noticeable difference in numbers between the lead and iron pipes (Kirmeyer et al, 2001).

Goethite and magnetite are common types of corrosion products found on iron piping. The presence of goethite is of interest because it is capable of adsorbing organics from the bulk fluid (Gu, et al 1994). High molecular weight humic and fulvic acids contain functional groups that can react with oxide solids on the pipe surface. Studies have also shown that the adsorption of NOM onto ferric oxide surfaces is pH dependent, with a decrease in NOM adsorption observed with an increase in pH (Gu et al 1994; Tipping 1981). NOM adsorption can also be reduced by the presence of phosphates or sulfates in the water, with phosphate having a greater effect than sulfate (Gu et al 1994, Davis 1982). The adsorption of NOM onto oxide surfaces may impact biofilm growth by providing shelter and possibly nutrients to biofilms.

In summary, the corrosion of ferrous materials and the corresponding presence of tubercles on the pipe surfaces may impact regrowth in several ways:

1) by increasing the surface area of the pipe more sites for biofilm attachment will be available;
2) by providing cracks and crevices that protect bacteria from disinfection;
3) by exerting a chlorine demand on the system (LeChevallier, et al 1996) and;
4) by the ability of iron oxide corrosion products to adsorb organics from the bulk fluid, hence providing a higher concentration of carbon on the pipe surface than in the bulk fluid (van der Wende and Characklis 1990). The pits and nodules formed by corrosion may also concentrate nutrients and provide organisms with protection from water shear (Allen and Geldreich 1977; Victoreen, 1977, 1980, 1984).

Microbiologically Influenced Corrosion (MIC)

Bacterial biofilms have also been linked to increased corrosion in iron pipelines. This phenomenon is known as biocorrosion or microbiologically influenced corrosion (MIC). Until recently it was believed that MIC occurs mainly in anaerobic environments in the presence of sulfide-producing bacteria. However, more recent work has shown that several other types of microorganisms, including hydrogen-producing bacteria, iron bacteria, and aerobic bacteria, can play a role in MIC (Geesey, 1991). Iron bacteria have been found on pipe surfaces and in water samples in distribution systems in Southern California (Ridgway et al., 1981). Additionally, sulfate-reducing bacteria were detected in 80% of the tubercles in the Columbus, Ohio distribution system (Tuovinen and Hsu 1982). This evidence shows that bacteria implicated in MIC can be prevalent in drinking water distribution systems.

Iron-reducing bacteria contribute to corrosion through reduction reactions that dissolve the passive oxide/hydroxide layer on iron surfaces. With the destruction of this layer, the iron surface is re-exposed to the bulk fluid, allowing further corrosion to occur (Geesey et al., 2000). The presence of chloride ions can increase the electrochemical potential for corrosion by combining with the ferric ions produced by the iron-reducing bacteria to form ferric chloride [$FeCl_3$] (Geesey, 1991).

The chemical and metabolic differences between different types of bacteria in a biofilm can also contribute to corrosion. The varying exopolymers secreted by bacteria differ in their affinities for and interactions with metal ions. This can lead to the formation of a metal concentration cell in which areas underneath exopolymers, which have high affinities for the underlying metal, are anodic to those areas underneath exopolymers that have low affinities for the metal (Geesey et al., 1989).

Re-deposition

The presence of ferrous materials in the distribution system may also impact the presence of biofilm growth on other materials. Camper (1996) found that biofilm heterotroph and coliform numbers on plastic surfaces with redeposited iron are comparable to the parent corroded iron surfaces. Kirmeyer et al (2001) found deposits containing considerable oxidized iron on the interior surface of cement lined pipe, even though there was no exposed cast iron. The iron most likely was removed from pipe surfaces farther upstream and collected on the cement lining. The number of bacteria in this material was comparable to exposed iron surfaces in the same distribution system (Kirmeyer et al 2001).

Biofilms Impact on Corrosion

Not only does the presence of corrosion products on distribution systems materials impact biofilm growth, but the presence of biofilms can impact corrosion as well. The interactions between the biofilm and the pipe wall or scale material can affect corrosion in the following ways:

- by hindering the transport of chemical species necessary for the passivation of the metal surface
- by facilitating the removal of protective scale layers as biofilm detachment occurs

- by forming differential aeration cells as a result of uneven or patchy distribution of the biofilm
- by altering the gradients of oxygen, pH, and other components of the bulk liquid.

Biofilms can cause the formation of differential aeration cells on the pipe, where the tubercle exposed to the bulk water acts as a cathode because of continuous oxygen supply, and the bottom of the tubercle where there is no oxygen is the anode (Geesey 1991, Videla 1996). Biofilm accumulations may also form diffusion barriers for certain chemical species. The transport of oxygen to the metal surface may be impeded by this diffusion barrier or by the metabolic utilization of oxygen within the biofilm (Videla 1996). The outward diffusion of metabolites and/or corrosion products may also be impeded. If the bottom of the biofilm is anaerobic, corrosive sulfate reducing bacteria (SRB) can proliferate, even if there is a measurable dissolved oxygen concentration in the bulk water. Sulfate reducing bacteria reduce sulfate to sulfide and produce mixtures of sulfides and/or hydrogen sulfide, which can be corrosive to iron (Cragnolino and Tuovinen 1984). Hydrogen sulfide has been shown to cause reductive dissolution of hematite, goethite, and lepidocrocite, causing release of Fe^{2+} from the oxide surface (dos Santos Alfonso and Stumm 1992). Also, acid producing microorganisms may accumulate enough acid within the biofilm to decrease the pH dramatically (Little et al 1986).

Corrosion Control

Many studies have demonstrated that implementing corrosion control can decrease microbial growth on iron surfaces. One study showed that the addition of zinc-orthophosphate or polyphosphate can reduce biofilm densities in chlorinated iron annular reactors. The addition of zinc-orthophosphate also decreased biofilm densities in iron annular reactors treated with monochloramine (Abernathy, 1998). A survey of 31 North American water systems showed a link between the use of phosphate-based corrosion inhibitors and lower coliform levels (LeChevallier et al., 1996). Several factors have been suggested to play a role in the reduction of biofilms through corrosion control. Some of these factors include changes to the surface chemistry of the pipe surface, reduction in the leaching of ferrous iron from the pipe surface to increase disinfectant efficacy, decreased biofilm habitat, and reduced disinfectant demand of the pipe surface (Abernathy, 1998).

LeChevallier et al (1990a) observed that biofilms grown on iron pipe surfaces exerted a disinfectant demand as much as 10 times greater than biofilms grown on other pipe materials, indicating that biofilms on iron pipe surfaces were protected from inactivation by a free chlorine residual and that corrosion products may affect the disinfection efficiency of pipe surfaces. In more recent studies, pipe materials that were subject to corrosion (i.e. ferrous materials) were more likely to support coliform bacteria even when a disinfectant residual was maintained (Camper, 1996; LeChevallier et al 1996). LeChevallier et al (1998) found that biofilms formed rapidly on iron pipe surfaces, even with a 1-2 mg/L disinfectant residual. No difference in biofilm densities were noted between chlorine and chloramines. In the same study however, biofilm densities were lower when chloramines were applied versus chlorine for CPVC surfaces.

Disinfection efficiency has also been related to the use of phosphate corrosion control strategies. In a study by LeChevallier (1990a) the accumulation of corrosion products on iron pipes was shown to interfere with free chlorine disinfection of biofilms, however use of polyphosphate and zinc orthophosphate corrosion inhibitors was shown to improve this efficiency, with average decreases of biofilm viable counts of 100 and 34 fold, respectively.

LeChevallier also evaluated whether specific corrosion control operational guidelines could be established which would also improve disinfection of biofilms. The guidelines used were the Larson Index and corrosion rate measurements. LeChevallier found that disinfection efficiency was improved when the Larson index was < 1.0 (LeChevallier et al 1993). This study also evaluated several corrosion inhibitors (zinc orthophosphate, zinc polyphosphate, polyphosphate, and sodium silicate) and found that as long as low corrosion rates were obtained (<1 mpy), no inhibitor was more effective over another for biofilm control. These results were obtained from a model pipe system containing black iron pipe. This affect was also seen in the South Central Connecticut Regional Water Authority (SCCRWA) which had experienced elevated coliform levels in its distribution system for several years. Coliform bacteria had also been isolated from iron tubercles from pipes in the SCCRWA distribution system (Opheim et al 1988). After increasing the concentration of zinc metaphosphate (a polyphosphate), coliform levels decreased from 12.9% to 5.1% over the next two years. Turbidity levels also decreased (Smith, et al 1989). Abernathy and Camper (2001) also found that the use of phosphorus based corrosion inhibitors significantly reduced biofilm densities.

There may be some potential negative impacts to the use of phosphate inhibitors as well. The potential for post-precipitation of zinc flocs with the use of zinc containing inhibitors could provide an environment which would encourage growth of bacteria. LeChevallier et al (1987) enumerated 10^7 bacteria/mL from a zinc floc observed in samples flushed from a dead-end main. The post-precipitation of zinc solids could produce a free chlorine demand as well as provide particles for attached microbiological growth. In addition, the use of corrosion inhibitors containing phosphorus may, in theory, contribute to algal growth in open reservoirs. This can result in turbidity and taste and odor problems as well as the potential for bacterial regrowth, as the by-products of algal growth may serve as additional nutrient sources for bacteria.

Other corrosion control strategies have also been shown to impact bacterial growth in the distribution system. Martin et al (1982) evaluated coliform bacteria growth from tubercles scraped from distribution system pipes and found that pH influenced the survival of organisms. Organisms exposed to alkaline pH (pH = 9.0) conditions had much lower survival rates than organisms exposed to pH 7.3 to 7.9 water, suggesting that high pH levels were bactericidal. Martin et al hypothesized that addition of lime would be an effective method for bacterial control, and cited evidence from Halifax, Nova Scotia. In this system, coliform problems in the distribution system were alleviated with a pH elevation from a range of 6.8-7.2 to a range of 8.5-9.0. LeChevallier et al (1990a) have shown that pH and alkalinity adjustment improved the efficiency of chlorine disinfection towards biofilms on iron pipes in a model pipe system, however, the actual mechanism for the observed reductions is unknown.

SUMMARY

The term *regrowth* is used to describe the chronic or periodic appearance of bacteria in the distribution system, either in the bulk water or at the pipe walls in a biofilm. There are many factors that can impact regrowth in water distribution systems, including nutrient levels, presence of disinfectants, temperature, watershed dynamics, hydraulic regime and the type of material used in the distribution system. As discussed previously, a variety of materials may be found in water distribution systems, however ferrous materials constitute the greatest percentage of pipe in place. In many cases, iron pipe has been in place for several years and has been subjected to the effects of internal corrosion and the formation of corrosion by-products on the interior surfaces.

The corrosion of iron has been recognized as being one of the primary factors affecting biofilm growth (Volk, 2000). Studies suggest that biofilms on iron surfaces are protected from chlorine residuals due to the reaction of corrosion products with the free chlorine (LeChevallier et al., 1990a). In addition, the corrosion of iron pipe can produce tubercles, which increase the surface area of the pipe. Cracks and crevices can provide protection from hydraulic currents and disinfectants for biofilm growth. Corrosion is also linked to increased binding of organic compounds and increases the hydraulic mixing in the bulk fluid, allowing for better transportation of nutrients to the surface (LeChevallier et al., 1996). The combination of these factors creates an ideal situation for increased microbial growth.

Ferrous materials in particular have been found to support significantly higher biofilm densities than non-ferrous materials, and iron pipe surfaces may protect bioflims from disinfection. Heterotrophic bacteria, including coliforms, iron bacteria and sulfate reducing bacteria have all been found to be associated with iron tubercles. Iron corrosion products can also slough off and re-deposit on non-ferrous materials in the distribution system, providing a suitable environment for biofilm growth. The corrosion of iron materials in the distribution system and the formation of tubercles on pipe surfaces can impact regrowth by providing more sites for biofilm attachment, by protecting bacteria from disinfection, by exerting a chlorine demand on the system and by adsorbing organics and providing a higher concentration of carbon on the pipe surface for bacterial growth.

Based on previous research it is expected that biofilms grown on iron surfaces in the presence of NOM and humic substances and/or disinfectants will be most problematic, since an increase in the bioavailability of NOM and humic substances in the presence of iron and a decreased efficacy of chlorine or monochloramine in the presence of iron has been demonstrated.

Little research has been conducted on the relative tendencies of cement, epoxy, and PVC to support biofilm growth. Since these materials have not been shown to have any significant interactions with NOM and humic substances or disinfectants they are not expected to support as much growth as iron. The available literature also suggests that the PVC is likely to support less biofilm growth than the cement.

Previous studies have also shown that in general, monochloramine is better than chlorine at controlling biofilm growth. The same literature has shown that at low residuals (<0.5 mg/L) the efficacy of monochloramine is significantly reduced (Sanderson et al., 1997).

CHAPTER 3
CASE STUDIES OF REGROWTH EVENTS RELATED TO MATERIALS USE

INTRODUCTION

In addition to the laboratory and field components of this project, two surveys were conducted to more closely examine the role that infrastructure materials may have on regrowth. The approach for this portion of the project was to collect, evaluate, and summarize information on regrowth events from selected case study utilities. These utilities were selected based on a preliminary survey sent to 185 utilities throughout the U.S. From this initial information, a select number of utilities were chosen for more detailed, in-depth case study write-ups. Case study write-ups were completed for utilities that had experienced mild to severe coliform positive problems as well as those that had experienced no coliform problems. Materials of use, water quality, and operational procedures employed by the utility were documented. Both the preliminary and the in-depth survey results are described below, along with a discussion of the major findings.

PRELIMINARY SURVEY

A total of 185 utilities were contacted and sent an initial distribution system survey. The survey contained questions on general utility characteristics such as sources, population served, treatment processes and distribution system characteristics. In addition, questions on general water quality and coliform incidence were included (a copy of the survey can be seen in Appendix A). Sixty-one surveys were returned, a 32% return rate, however one utility declined to give any information other than the coliform incident information. All of the respondents were medium or larger systems (serving more than 10,000 people) that take 40 or more total coliform samples per month. Two thirds of the respondents were located east of the Mississippi River, and these eastern utilities generally had higher levels of unlined cast iron piping in their distribution systems. Figure 3.1 displays the breakdown of distribution system pipe age for these utilities, and Figure 3.2 show the percentage of unlined cast iron piping.

Within these 61 responding utilities, there was a good spectrum of coliform positive patterns. The breakdown was as follows:

20 utilities (32.8%) – no coliform events
18 utilities (29.5%) – coliform positives at 2.5% or less
11 utilities (18%) – coliform positives at 5% or less
12 utilities (19.7%) – coliform positives at more then 5%

Table 3.1 displays a more detailed breakdown of the utilities that responded to the preliminary survey.

The majority of the responding utilities experienced either no coliform incidents over the past 10 years, or incidences at a low level of detection, i.e. 0 – 2.5%. Five utilities reported that they had experienced acute coliform events, i.e. 5% exceedance levels and five utilities had coliform presence at all levels over long periods, which could be indicative of regrowth. Of the eighteen utilities that reported coliform incidences in the range of 0 to 2.5%, six utilities had chronic low levels of coliform presence. Chronic low level coliform presence was defined as several years of reporting incidences of 2.5% or less more than four times in any given year.

Table 3.1
Initial Survey Respondents

	Sources of Water				
	Surface	Ground	Both	Purchased	Total
Number of Systems	32 (53.3%)	9 (15%)	17 (28.3%)	2 (3.3%)	60 (100%)
<u>Coliform Percent Incidence Ranges</u>					
No Coliform Incidents	10	5	5	0	20
Coliform at 0 – 2.5% only	11	2	4	1	18
Coliform at 2.5 – 5% only	2	0	0	1	3
Coliform >5% Only	3	1	1	0	5
Coliform at 0 – 2.5%, and 2.5 – 5%	3	1	5	0	9
Coliform in All Ranges	3	0	2	0	5

The utilities that reported acute coliform exceedences of greater than 5% attributed these coliform events to storm events, contamination during main breaks or replacement, and flow reversals in the distribution system. Four of these five systems have unfiltered surface water sources.

Utilities with chronic coliform incidences (>2.5%) commonly had chlorine residuals less than 1 mg/L in the distribution system, and this was cited as a cause of the coliform positives by three of the utilities. Most of these incidences had no seasonal trend, but one system did report incidences during warm weather months primarily. All six of the utilities defined as experiencing chronic low level coliform occurrence filtered their source water. Four were surface water sources, one was groundwater, and one had both surface and groundwater sources.

For the utilities that reported coliform in all percent incidence ranges, three of the six were unfiltered surface water sources, one was a mix of filtered and unfiltered, two were groundwater and surface water with filtration. Low chlorine residuals were sited by all six of these utilities as a causative factor, and in general, coliform levels increased with increases in water temperature

IN-DEPTH CASE STUDIES

Approach

Ten utilities were chosen for a follow up survey. The utilities chosen represented three different ranges of coliform incidence; 1) the chronic, high level coliform pool (> 2.5%), 2) low level (< 2.5%) pool, and 3) the zero incidence pool. Table 3.2 lists the individual utilities.

Relative numerical scoring factors were assigned to each of these sub-categories, as seen in Table 3.3. For example, for the coliform incidence category, utilities with coliform incidences of> 5% were given assigned a value of 1, whereas utilities with a coliform incidence of zero were assigned a value of 4. These numerical scores are arbitrary, and were applied to the information gathered from each in-depth case study utility only to provide a standard method for assessing the potential correlations each of these sub-categories my have on coliform occurrence.

Table 3.2
In-depth Case Study Utilities

Coliform Incidence Range	Utility	Source Characteristics	Distribution System Characteristics
0% Coliform	Water District No. 1 of Johnson County, Kansas City, KS	Softened surface water	5% unlined cast iron pipe
	Richland, WA	Direct filtered surface water	6% unlined cast iron pipe
0 – 2.5% Coliform	Columbus, OH	Filtered surface water	40% unlined cast iron pipe
	Carrboro, NC	Filtered surface water	19% unlined cast iron pipe
	Gilbert, AZ	Mixed groundwater and filtered surface water	0% unlined cast iron pipe
	Kennewick, WA	Mixed groundwater and filtered surface water	15% unlined cast iron pipe
> 2.5% coliform	Greenville, SC	Mix of filtered and unfiltered surface water	20% unlined cast iron pipe
	Scottsdale, AZ	Mixed surface and ground water	No unlined cast iron pipe
	Boston Water and Sewer, MA	Unfiltered surface water	55% unlined cast iron pipe
	Revere, MA	Unfiltered surface water	70% unlined cast iron pipe

Each of the utilities listed in Table 3.2 were sent a detailed survey form, which can be found in Appendix A. This survey contained questions on water quality, treatment changes, tracking of customer complaints, problem sample locations, water ages, flushing practices, maintenance and construction, breaks, storage facilities, and modeling.

The information gathered from the ten utilities was categorized into three major areas, in order to evaluate relative effects on coliform incidence. The first major area was infrastructure, the second major area was distribution system operations and maintenance practice, and the third major area was water treatment and quality. The subcategories of information gathered from each utility for each of these three major areas were as follows:

Infrastructure	Distribution Operations and Maintenance	Water Quality and Treatment
Average pipe age	System flushing	Coliform incidence
Oldest pipe age	Replace/repair	Chlorine/chloramine residuals
% Cast iron pipe	Tank cleaning	pH
Storage turnover		Filtration
		Proportion of supply that is ground water

The information obtained from each utility in each of these subcategories was scored using the values shown in Table 3.3. Discussions of the results of this evaluation matrix are provided below, and the case study write-ups for theses 10 utilities are in Appendix B.

Findings

Infrastructure

The system's infrastructure may play a significant role in the potential for coliform incidence. Systems with older pipe, i.e., average pipe age greater than 40 years and maximum pipe age older than 80 years were associated with higher coliform incidence. Long residence times (greater than 3 days) or large amounts of distribution system storage (greater than 2 days) seemed to increase the chance of coliform incidence. In general, higher proportions of unlined cast iron pipe within a distribution system were associated with higher incidence of coliform, though there were exceptions to this observation (Columbus, Ohio). Table 3.4 summarizes the infrastructure evaluation for the 10 utilities.

Table 3.3
Relative Numerical Scoring Factors

Major Areas	Sub-category	Best=4	Good=3	Ave=2	Poor = 1
Water Quality and Treatment	Coliform	0	<2.5%	2.5-5%	>5%
	Chlorine/chloramine residual	Residuals listed as actual concentration. Chlorine given a weighting factor of 5			
	PH	>9	8-8.5	7.5-8	7-7.5
	Percent of source filtered	100%			0%
	Proportion of groundwater	GW (groundwater) is actual proportion of source water from groundwater			
Infrastructure	Pipe age (aver)	<20	20-40	41-60	>60
	Pipe age (oldest)	<40	41-60	61-80	>80
	Percentage of cast iron pipe	<10%	10-25	26-50	>50%
Distribution System Operations and Maintenance	Storage	<0.75 day	.75-1.25	1.26-2 day	>2 day
	Flush system	100%	75-50%	49-25%	<25%
	Replace/repair	Disinfect Flush Test	Disinfect Flush	Flush	Nothing
	Tank cleaning	1 year	2 year	3-5 year	>5 year

Table 3.4
Infrastructure Evaluation

Location	Infrastructure Scores[1]				
	Average Pipe Age	Oldest Pipe Age	Percentage of Cast Iron Pipe	Storage Turnover	Sum
Richland, WA	3	3	4	3	13
Water District #1	3	2	4	3	12
Columbus, OH	3	1	2	3	9
Gilbert, AZ	4	4	4	2	14
Carrboro, NC	3	2	3	3	11
Kennewick, WA	3	2	3	1	9
Scottsdale, AZ	4	2	4	3	13
Revere, MA	1	1	1	1	4
Boston, MA	1	1	1	1	4
Greenville, SC	2	1	3	2	8

[1] 4=Best, 3=Good, 2=Average, 1=Poor

Distribution System Operations and Maintenance

Distribution system operation and maintenance or, more accurately, lack thereof also contributed to coliform incidence. Frequent cleaning and inspection of distribution system storage facilities in cycles of two years or less was associated with lower coliform incidence. A few utilities initiated annual inspection and cleaning in response to Coliform Rule exceedance and found a decrease in coliform incidence as a result. Pipe repair and replacement practices were also important. Those utilities that reported practices of flushing, disinfecting and coliform testing of all pipe repairs and pipe replacement activities were found to have the lowest coliform incidence. Utilities that reported only two of the three listed practices tended to have higher coliform incidence.

Few of the utilities accomplished complete flushing of the system every year. Most flushed along dead ends or in known problem areas. However, once a coliform or regrowth problem developed, system wide, unidirectional flushing was an important step in restoring water quality and has proved beneficial to all of the utilities with high incidence of coliform (Boston, Revere, Greenville). Table 3.5 summarizes the operations and maintenance evaluation for the ten utilities.

Table 3.5
Operations and Maintenance Evaluation

	Operations and Maintenance Scores[1]			
Location	System flushing	Tank Cleaning	Replace/Repair	Sum
Richland, WA	1	3	4	8
Water District #1	2	2	2	6
Columbus, OH	1	3	3	7
Gilbert, AZ	1	1	3	5
Carrboro, NC	2	4	3	9
Kennewick, WA	2	4	3	9
Scottsdale, AZ	1	4	3	8
Revere, MA	3	3	3	9
Boston, MA	1	4	2	7
Greenville, SC	1	1	3	5

[1] 4=Best, 3=Good, 2=Average, 1=Poor

Water Quality and Treatment

Chlorine residual levels, either as free chlorine or total chlorine in those systems using chloramines, were strongly associated with coliform incidence. Free chlorine residuals at less then 1 mg/L entering the distribution system, or total chlorine residuals at less than 2.5 mg/L in warm water locations and 1.5 mg/l at cooler water locations, tended to predispose the system to coliform incidence.

Water source and treatment of that source also appeared to be associated with coliform incidence. Systems supplied entirely by surface water sources, particularly unfiltered surface water sources, were at greater risk of coliform incidence.

The pH of the water within the distribution system also appeared to play a role in coliform incidence, possibly facilitating survival at the lower pH values (pH 7.5 or less). High pH values (greater than pH 9) in the distribution system were associated with lower incidences of coliform. As most enteric and environmental bacteria normally live in near neutral or slightly acidic pH environments, it would make some sense that higher pH environments would be hostile to coliform survival, even though disinfectants may not be as effective at the higher pH values. Table 3.6 summarizes the water quality and treatment evaluation for these ten utilities.

Table 3.6
Water Quality and Treatment Evaluation

Location	Water Quality and Treatment Scores						
	Coliform Incidence[1]	Chlorine Residual[2]	Chloramine Residual[3]	PH[1]	Percent of Source Filtered[1]	Proportion of Groundwater[4]	Sum
Richland, WA	4	5		1	4	0.1	10.1
Water District #1	4		2.7	4	4	0.03	10.73
Columbus, OH	3	7.5		2	4	0.11	13.61
Gilbert, AZ	3	6.5		2	4	0.48	12.98
Carrboro, NC	3	5		1	4	0	10
Kennewick, WA	3	5		2	4	0.6	11.6
Scottsdale, AZ	2	5		2	4	0.3	11.3
Revere, MA	1		1	1	1	0	3
Boston, MA	1		0.5	1	1	0	2.5
Greenville, SC	1		3	1	2.5	0	6.5

[1] 4=Best, 3=Good, 2=Average, 1=Poor; [2] Actual Average Chlorine residual x 5 (weighting factor); [3]Actual average chloramine residual; [4] Actual proportion of source water that is groundwater

SUMMARY

Table 3.7 contains a summary of the overall results for the 10 utilities. Less than optimal water quality and infrastructure appeared to be the most important components predisposing water systems toward coliform incidents or regrowth. Those systems that scored less than 10 in either water quality or infrastructure categories had coliform incidences at greater than 5%. Operation and maintenance played a lesser role and the spread of data between utilities in this area was less than five points, with no utility scoring above 9 in this area.

Table 3.7
Summary of Scores

Location	Coliform Activity	Scores			
		Infrastructure	Operations and Maintenance	Water Quality and Treatment	Total
Richland, WA	0	13	8	10.1	31.1
Water District #1	0	12	6	10.73	28.73
Columbus, OH	<2.5%	9	7	13.61	29.61
Gilbert, AZ	<2.5%	14	5	12.98	31.98
Carrboro, NC	<2.5%	11	9	10	30
Kennewick, WA	<2.5%	9	9	11.6	29.6
Scottsdale, AZ	2.5% - 5%	13	8	11.3	32.3
Revere, MA	>5%	4	9	3	16
Boston, MA	>5%	4	7	2.5	13.5
Greenville, SC	>5%	8	5	6.5	19.5

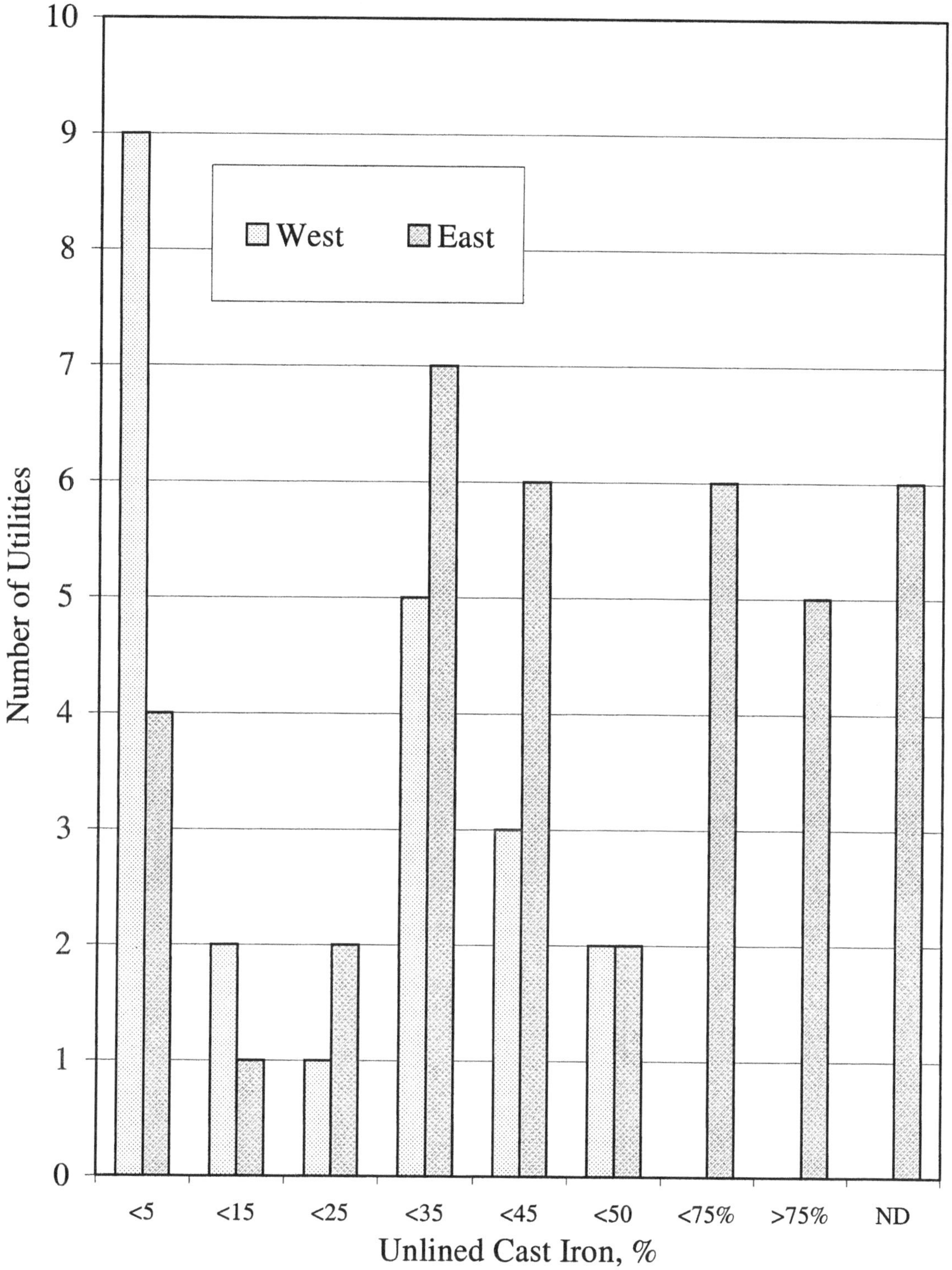

Figure 3.1 Pipe age distribution for survey utilities

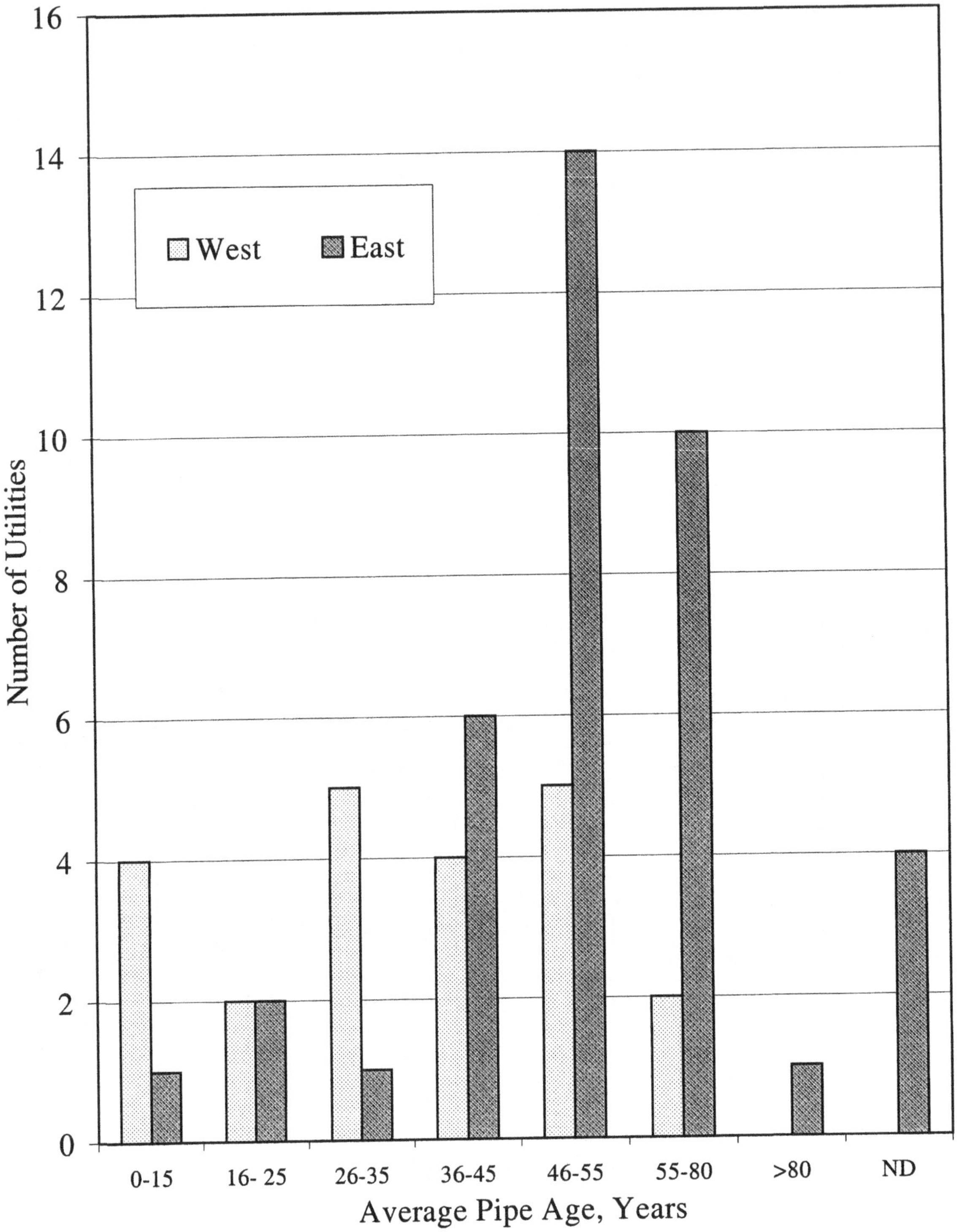

Figure 3.2 Percent unlined cast iron pipe

CHAPTER 4
LABORATORY STUDIES OF MATERIALS AND REGROWTH

INTRODUCTION

Laboratory experiments were conducted at Montana State University and run in parallel with the field studies described in Chapter 5. A series of experiments were developed to evaluate the relative importance of two key water quality parameters; 1) the amount of biodegradable organic matter, and 2) the type of disinfectant used, as they relate to biofilm accumulation on unlined ductile iron, PVC, epoxy, and cement surfaces. These experiments, along with a discussion of the results, are described below.

METHODS AND MATERIALS

Experimental Setup

The laboratory setup consisted of four pairs of annular reactors. Each pair of reactors contained coupons of only one of the following materials: epoxy, ductile iron, cement, or PVC. The annular reactors used in this experiment were model 920LJ manufactured by Biosurface Technologies Corporation. These reactors have a variable speed rotating drum, a volume of roughly 1 liter, and a high ratio of surface area for growth to the fluid volume. Each annular reactor drum holds 20 coupons measuring roughly 1 cm in width and 15 cm in length. The rotational speed of the reactors was set at roughly 90 rpm to simulate the shear stress in a four-inch pipe with a fluid velocity of 1 ft/s. This rotational speed was selected based on calculations provided by the manufacturer of the annular reactors. A schematic diagram of the annular reactor design can be seen in Appendix C.

Bozeman tap water flowed through a granular activated carbon (GAC) column and then through a biologically activated carbon (BAC) column into a 2 liter holding tank. The columns were operated in an up-flow mode. The purpose of using the GAC and BAC columns was to remove chlorine and some organic carbon from the Bozeman tap water and create an influent of consistent biological and chemical quality to the annular reactors. From the holding tank this water was pumped at a constant flow rate into the annular reactors to achieve a detention time of 2 hours per reactor. Block et al (1993) found that the growth rate of HPC in an experimental drinking water distribution system was 0.0017/hr with no chlorine at the inlet where BDOC was highest. This correlates to a doubling time significantly higher than 2 hr (17 hr). These authors also state that the doubling time of bacteria in distribution systems is expected to exceed 100 hours. Van der Wende, Characklis and Smith (1989) reported HPC biofilm doubling times of approximately 17 days in annular reactor systems fed unchlorinated tap water, with negative growth rates for suspended cells. These same authors state that for the number of cells that appeared in the reactor effluent to arise only from replication of planktonic cells, their growth rate would need to be unrealistically high, at 0.5/hr. In addition, Haudidier et al. (1988) calculated that planktonic growth rates are very low or negative in distribution systems. Therefore, a 2 hour detention time was deemed sufficient to allow biofilm growth while minimizing planktonic growth four these studies.

The laboratory experiment was divided into four phases. Each phase lasted a minimum of 3-4 months to allow the processes in the annular reactors to approach steady state. Sampling was initiated approximately one week after the start of each phase. In each phase all reactors were fed with the Bozeman tap water as described above, and a nitrogen/phosphate solution. The nitrogen/phosphate solution was added to ensure that the reactors were carbon limited. In the first control phase nothing else was added to the reactors. In the second phase chlorine was added to one reactor in each pair and monochloramine was added to the other reactor to achieve target effluent concentrations of 0.2 mg/L as free chlorine and total chlorine, respectively. In the third phase humics-derived carbon was added to all reactors at a concentration of 0.5 mg/L TOC. In the fourth phase the humics-derived carbon was increased to 2 mg/L TOC. In phases three and four chlorine and monochloramine were also added to achieve the target effluent concentration of 0.2 mg/L as free chlorine and total chlorine, respectively. Table 4.1 summarizes the water quality conditions for each phase of the experiment.

In phase one fresh coupons were used in all 20 slots on the reactor drum. At the beginning of phase two, alternate coupons on the drum were removed and replaced with new coupons. At the beginning of phase three, all coupons that were not removed at the beginning of phase two were replaced with fresh coupons. At the beginning of phase four, the same coupons that were replaced at the beginning of phase two were again replaced with fresh coupons. Thus in each of the last three runs 10 new coupons were available for examination of initial growth of new biofilms. In addition, 10 coupons containing biofilms from the previous phase were available for the examination of the less dramatic changes in older biofilms nearer to equilibrium. New coupons and old coupons were sampled in alternate weeks starting with the new coupons.

Table 4.1
Water Quality Conditions for the Four Phases of the Laboratory Experiment

Phase	GAC/BAC Water	Nitrogen/ Phosphorus	Disinfectant [1]	Humics-Derived Carbon
1 (Control)	Yes	Yes	No	No
2	Yes	Yes	4 with Chlorine 4 with Monochloramine	No
3	Yes	Yes	4 with Chlorine 4 with Monochloramine	0.5 mg/L TOC
4	Yes	Yes	4 with Chlorine 4 with Monochloramine	2 mg/L TOC

[1] For the phases where disinfectant was added, chlorine and monochloramine were added to achieve target effluent concentrations of 0.2 mg/L as free chlorine and total chlorine, respectively

Chemical Solutions Preparations

Chlorine and Monochloramine Preparation

The chlorine feed solution was prepared from household bleach (sodium hypochlorite) containing no additives or buffers. A small volume of bleach was injected into a large volume of water. The appropriate ratio of water to bleach for each reactor was determined through trial and error to achieve roughly a 0.2 mg/L residual of free chlorine in the reactors. The ratio changed frequently during the course of the experiment to accommodate fluctuations in the water quality in the annular reactors. The chlorine feed was always operated at a flow rate of approximately 0.25 mL/min into each treated reactor.

Monochloramine was prepared using techniques developed in previous studies at the Center for Biofilm Engineering. A phosphate buffer was made by adding 0.5 g of dibasic potassium phosphate to 1 L of ultrapure water in a sterile glass bottle. The pH of this buffer was then adjusted to between 8.9 and 9.2 using 0.1 N NaOH. In a separate container, 0.11 g of ammonium chloride was added to 100 mL of the phosphate buffer and mixed. While continuing to mix this solution, 1 mL of a ~4% sodium hypochlorite bleach solution was added at a rate of no more than 20 μL per 6 seconds. The solution was allowed to stir for 30 minutes before the chlorine residuals were measured.

The concentration of the monochloramine was checked for consistency with the HACH DR/2000 using DPD powder packets and the HACH method 80. The difference between the total and free chlorine residuals was considered the monochloramine concentration. The method described above for monochloramine preparation produced a concentration of 400-500 mg/L.

The appropriate level of dilution of the monochloramine stock solution for each reactor was determined through trial and error to achieve roughly a 0.2 mg/L residual of total chlorine in the reactors. The level of dilution changed frequently during the course of the experiment to accommodate fluctuations in the water quality in the annular reactors. The monochloramine feed was always operated at a flow rate of approximately 0.25 mL/min into each treated reactor.

Nitrogen/Phosphate Solution Preparation

A molar ratio of 100:10:1 for carbon:nitrogen:phosphorus was used to determine the nitrogen and phosphorus needs in the annular reactors. This calculation was based on the average carbon levels measured in the reactors and was adjusted only when significant fluctuations were observed and at the beginning of each phase when the carbon was deliberately increased. Potassium nitrate was used as the nitrogen source and the phosphorus was derived from equimolar concentrations of dibasic and monobasic potassium phosphate. A stock solution was created such that one mL of stock solution was added per L of autoclaved ultrapure water in the feed solution. The flow rate of this feed solution into each reactor was approximately 0.25 mL/min.

Humics Solution Preparation

The humics solution was prepared using Elliot Silt Loam obtained from the International Humic Substances Society and has been used extensively in research projects at the Center for Biofilm Engineering. 100g of this soil was added to 1L of 0.1 N sodium hydroxide in a baked glass bottle to minimize carbon contamination. This soil solution underwent constant mixing for 2 – 4 days after which it was centrifuged at a g-factor of 4 for 20 minutes. The supernatant was then poured into another baked glass bottle and a small sample was removed and diluted with nanopure water for determination of the carbon concentration of this stock solution.

The dissolved organic carbon (DOC) concentration of the solution was measured with the Shimadzu TOC-5000A Total Organic Carbon Analyzer using the methods described below for the annular reactor samples. The final concentration usually measured between 1000 mg/L and 1500 mg/L DOC depending on the length of the mixing time. The volume of humics stock solution to add to the feed jug was calculated using the flow rate and volume in the annular reactors, the volume of autoclaved ultrapure water in the feed jug, and the desired concentration of humics to be added to the reactors. After the humics were added to the ultrapure water in the feed jug, the pH was adjusted with 2N hydrochloric acid (HCl) to match the pH in the influent as closely as possible. The flow rate of the humics feed into the reactors was approximately 0.25 mL/min. The amount that is biodegradable was determined to be 28% of the total organic carbon, which was determined in a separate study using a re-circulating BDOC column (Buls, 2000).

Biological Analyses

Heterotrophic Plate Counts

One coupon was removed for sampling from each reactor roughly once every week during each phase, beginning one week after the initial startup. The surface of the coupon was scraped using a flat-headed spatula into a 100 mL beaker with 10 mL of sterilized water. The contents of the beaker were then carefully transferred to a sterile test tube for further handling. The influent water was sampled by taking 10 mL from the holding tank and placing it in a sterile test tube. The reactor effluent lines had a tendency to grow biofilms along the sides of the tubing which would detach during sampling so effluent samples of 10 mL were instead drawn directly from the inside of the annular reactors. Since the fluid in the annular reactors is thoroughly mixed the concentrations in the effluents are the same as the concentrations inside the reactors.

R2A plates were used for the heterotrophic plate counts. These plates were prepared using 12.7 g of a pre-mixed powder, which was added to 700 mL of water and autoclaved for a minimum of 30 minutes. The autoclaved media was allowed to cool in a water bath for 45 minutes then poured into petri dishes. These R2A plates were allowed to sit for a minimum of two days before use to allow some drying to occur and to ensure that the plates were not biologically contaminated during pouring.

Biofilm samples from the reactors were homogenized using a Janke & Kunkel model T 25 S1 homogenizer for 30-60 seconds at 20,500 min^{-1} to break up clumps of the biofilms before they were diluted for plating. The homogenizer probe was soaked in alcohol, which was burned off, then rinsed in sterile dilution water between samples to reduce the risk of cross contamination. After homogenization, the biofilm, influent, and effluent samples were diluted one or more times by transferring 1 mL to a sterile test tube containing 9 mL of water using aseptic techniques. This process was repeated for each sample until an appropriate final dilution was achieved for that sample. Then three 0.1 mL samples were drawn from each dilution and spread on three separate R2A plates using aseptic techniques. All R2A plates containing samples were then incubated at room temperature for 7 days before the colonies on each were counted. Plates with the sterile dilution water and the rinse water from the homogenizer probe were also incubated as controls. After all samples had been plated, 0.2 mL of filter sterilized 33% formaldehyde was added to each sample tube. Tubes were refrigerated for a minimum of 24 hours to preserve the samples for total direct counts.

The average number of colony forming units per mL of sample was calculated based on the average number of colonies per plate, the dilution plated, and the volume plated. For the biofilm samples this number was converted to the average colony forming units per cm^2 of the coupon using the area of the coupon and the volume of sterile dilution water that the biofilms were scraped into.

Total Direct Counts

Samples for the total direct counts were vortexed for 1 minute to ensure thorough mixing, then vacuum filtered through 0.22 μm polycarbonate filters. The polycarbonate filters used in this study were black, 25 mm diameter filters manufactured by Osmonics, Inc. (material #1215609). 0.25 to 0.5 mL of 100 mg/L DAPI (4',6-diamidino-2-phenylindole) stain from Sigma was added and allowed to stand on the filter for 15-20 minutes before being vacuum filtered. The filter was then fixed onto a microscope slide and cells in 20 randomly selected fields from the filter were counted under a UV light on the Nikon Eclipse E800 microscope. The average number of cells per field were calculated and used to calculate the average number of cells per filter based on the area of the counting field and the area through which the sample was filtered. The cells were counted with a 100x objective with a 10x/25 eyepiece.

Chemical Analyses

Total Organic Carbon

Glassware and glass sample vials for the TOC analyses were soaked in a 36 normal sulfuric acid bath for no less than 8 hours, then rinsed repeatedly with ultrapure water and covered with aluminum foil. The glassware and sample vials were then baked for 48 hours at roughly 350°C. This process cleaned and sterilized the glassware and sample vials and ensured minimal carbon contamination. Volumetric flasks and graduated cylinders did not undergo baking as the high temperature tends to slightly warp glassware, which would affect the accuracy of the volume markings.

Samples for TOC analysis were filtered through Fisherbrand 0.2 μm nylon filters. A cleaning process was devised in prior research projects to minimize carbon contamination from these filters. The filters were rinsed with 30 mL of 0.1 N sodium hydroxide 3 times followed by 30 mL of ultrapure water 3 times. A fourth rinse with ultrapure water immediately preceded sampling and the filtrate was measured to ensure that the background carbon levels on the filters had been reduced to 100 ppb or less.

Samples were collected by inserting the tip of a sterile 30 mL syringe directly into the annular reactors or water holding tank. These samples were each filtered through the cleaned filters described above into the acid washed and baked sample vials. The samples were acidified with 0.2 mL of 2 N hydrochloric acid and stored at 4°C until measurement. TOC was measured as non-purgeable organic carbon on the Shimadzu TOC-5000A Total Organic Carbon Analyzer, using a high temperature catalytic method with a high sensitivity catalyst for low carbon analysis. Samples were sparged for 5 minutes during measurement to remove all inorganic and volatile organic carbon.

Prior to sample measurement of TOC, a standard curve was developed using potassium hydrogen phthalate ($KC_8H_5O_4$), which was baked in an oven for at least one week until dehydration was achieved. 1.0626 g was then added to 1.0 L of ultrapure water in an acid washed and baked volumetric flask to create a stock solution with a concentration of 1000 mg/L C. 25 mL of this stock solution was then added to 1.0 L of ultrapure water to create a second stock solution with a concentration of 25 mg/L. This stock solution was then further diluted in separate volumetric flasks to obtain concentrations of 250, 500, 1000, 2000, and 4000 µg/L C. A four point linear regression was performed using four of these concentrations to calibrate the Shimadzu for sample analysis. This calibration was performed every time new standard solutions were made or instrument maintenance took place.

Free/Total Chlorine

Free and total chlorine measurements were made using HACH method 80 with the HACH DR/2000 and DPD powder packets specific to each test. Reactors containing monochloramine as a disinfectant were maintained at 0.2 mg/L as total chlorine. Reactors containing chlorine as a disinfectant were maintained at 0.2 mg/L as free chlorine. Measurements were taken in each reactor every 2-3 days during phases 2, 3, and 4 to ensure that these target levels were achieved. When significant deviations from these target levels were noticed, the monochloramine or chlorine concentrations in the feed jugs were increased or decreased accordingly.

Bozeman Water Quality Data

Table 4.2 lists the water quality data for the finished water from the Bozeman drinking water treatment plant for the year 2000. Bozeman tap water was passed through GAC and BAC columns and then used as a source water in this study.

Statistical Analyses

A one-way analysis of variance (ANOVA) was performed using MINITAB$_{TM}$ Version 13.2. The logarithms of the cell densities (either total direct counts or HPC levels) were entered as the response variable. All confidence intervals calculated as part of the one-way ANOVA were simultaneously correct with probability 0.95. The associated significance tests have a simultaneous Type I error rate of 0.05. The Bonferroni method was used to determine the individual error rate to be used in the statistical analyses; it divided the desired simultaneous error rate of 0.05 by the number of pairwise comparisons that were considered in each analysis. For example, when the effluents from the four reactors containing different materials were compared to each other in a given phase with a given disinfectant there were a total of six pairwise comparisons (6 equals the number of combinations of 4 things taking 2 at a time or 4C2 on the typical electronic calculator). Thus, the individual error rate was 0.0083 and the simultaneous error rate for the analysis was only 5%.

Table 4.2
Bozeman Water Quality Data for the Year 2000

Parameter	MCL	Units	Range	Average
Alkalinity	N/A	ppm	50 - 103.8	83.7
Chlorine residual, free (>0.2)	4	ppm	1.2 - 1.9	1.5
Flouride (add to = 1.0)	4	ppm	0.0 - 1.4	1.0
Hardness, calcium	N/A	ppm	38 - 73	59.3
Calcium	N/A	ppm	15.2 - 29.2	23.7
Hardness, magnesium	N/A	ppm	17.5 - 42.8	29.5
Magnesium	N/A	ppm	4.3 - 10.5	7.4
Hardness, total	N/A	ppm	56.8 - 110.4	88.7
Hardness, total	N/A	Grain/gal	3.3 - 6.5	5.2
PH	6.5-9.3	units	7.4 - 9.0	8.5
Sodium	20	ppm	1.3 - 7.9	3.5
Sulfate	500	ppm	0.0 - 9.3	0.95
Iron	0.3	ppm	0.01 - 0.21	0.04
Total dissolved soils	500	ppm	56.2 – 111.4	92.4
Turbidity (daily average)	0.5	NTU	0.03 – 0.18	0.06
Total coliforms	There were no positive samples in Bozeman's drinking water after treatment.			

Much of the data described in the following sections are displayed graphically using boxplots. These boxplots consist of three major features; a box, a horizontal line drawn through the box, a pair of vertical lines extending from the upper and lower ends of each box, and asterisks. The box portion of the plot represents the middle 50% of the observations. The line drawn through the box represents the median of the data. The lines extending from the box are called whiskers. These whiskers indicate the lowest and highest values in the data set (excluding outliers). The asterisks represent possible outliers in the data set. A data point is considered an outlier if it is outside of the box by more than 1.5 times the middle 50% of the observations.

RESULTS

Total Organic Carbon Analyses

Table 4.3 shows the average, high, and low values for the dissolved organic carbon (DOC) measured in each reactor and the influent in each phase of the experiment. This table shows a visible increase in the DOC in all reactors between phases two and three and phases three and four. This corresponds to the addition of supplementary humic derived carbon in phases three and four. Table 4.4 shows that this increase was primarily due to the supplementary carbon and not to increases in the Bozeman tap water carbon concentration. In the first two phases, where no supplementary carbon was added, the effluent DOC concentrations are all nearly equal to or below the level in the influent. In phases three and four all of the effluent DOC concentrations were greater than the influent DOC concentration.

Table 4.4 also indicates the relative carbon depletion in the reactors. The only carbon entering the reactors came from the influent and the supplementary humic substances feed, which were fed to all reactors at the same mass flow rate. Since biomass was filtered out in the sampling process, differences in the values of the effluent DOC minus the influent DOC indicate differences in carbon depletion within the reactors. Lower values indicate greater carbon depletion in a reactor. In phase one, the greatest carbon utilization occurred in the reactors containing PVC, followed by the reactors containing iron. In phase one, the greatest carbon utilization occurred in the reactors containing PVC, followed by the reactors containing iron. In phase one there was little difference between the carbon utilization in the epoxy and cement reactors. In all of the remaining phases the iron reactors utilized the most carbon. In phases two through four the order of ascendance varied among reactors containing materials other than iron.

Influent/Effluent Heterotrophic Plate Counts

There was no significant difference between the mean influent HPCs in any of the phases, indicating that fluctuations in the influent HPC concentrations were not a major factor influencing changes in the planktonic populations within the reactors between phases in the reactors. Thus, any variation in these populations can be attributed to experimental changes in the disinfectant and/or carbon levels in the reactors. The influent and effluent HPC data for each reactor type for phase one through four are illustrated in Figures 4.1 through 4.4. Statistical analyses of the effluent HPCs in the control phase (phase one) showed that there were no significant differences between reactor pairs containing the same materials, indicating that the results from these reactors are repeatable (data not shown). In addition, it was possible to use the average for each material in further statistical comparisons rather than making comparisons to individual reactors.

Statistical comparisons of the effluent cell concentrations for different materials for phases one through four are shown in Table 4.5. These results indicated that only the iron and cement reactors did not significantly differ from each other in phase one. Also shown in this table is that the effluent counts did not differ by more than one logarithm (10^4 cfu/mL). Another analysis showed that all of the effluent concentrations were significantly higher than the influent concentration (average of 4.01 log (cfu/mL)).

Table 4.3
Influent and Effluent DOC levels for All Reactors in All Phases, µg/L C

Phase	Humics derived Carbon added	Influent	Epoxy 1	Epoxy 2	Iron 1	Iron 2	Cement 1	Cement 2	PVC 1	PVC 2
1	None	912 (429-1602)	883 (418-1571)	884 (411-1593)	800 (355-1423)	790 (355-1389)	892 (432-1618)	915 (440-1593)	643 (289-1277)	631 (260-1314)
2	None	737 (645-868)	745 (629-965)	699 (610-835)	581 (426-703)	565 (491-671)	754 (658-878)	704 (616-876)	712 (618-871)	735 (652-841)
3	0.5 mg/L	619 (512-784)	1198 (572-4245)	1059 (561-4114)	745 (532-929)	638 (514-782)	1010 (732-1184)	961 (475-2558)	849 (539-1161)	873 (525-1059)
4	2 mg/L	1573 (1214-2482)	2573 (1960-3548)	2556 (1899-3584)	2080 (1163-3252)	1793 (1124-2642)	3119 (2394-4856)	2553 (1420-3766)	2840 (1459-4342)	2711 (1619-4445)

Table 4.4
Average Effluent DOC Minus the Average Influent DOC for All Reactors in All Phases, µg/L C

Phase	Humics derived Carbon added	Epoxy 1	Epoxy 2	Iron 1	Iron 2	Cement 1	Cement 2	PVC 1	PVC 2
1	None	-30	-28	-112	-122	-20	3	-269	-281
2	None	8	-38	-156	-172	17	-33	-26	-2
3	0.5 mg/L	579	440	127	19	391	343	230	254
4	2 mg/L	1001	983	507	220	1546	980	1268	1138

In phase two, no significant differences were found between the effluents from reactors containing the same materials that were treated with monochloramine or chlorine (data not shown). Only two similarities were observed between reactors treated with monochloramine and those treated with chlorine. The first was that the reactors containing iron had significantly higher microbial counts in the effluents than those containing PVC, which were approximately one logarithm (log) lower. The second was that the reactors containing cement did not have significantly different effluent counts than those containing PVC.

Statistical comparisons of the reactor effluents to the influent in phase two where disinfectants were added showed that there was no significant difference between the influent (4.18 log (cfu/mL)) and the effluent from the iron reactors treated with either monochloramine or chlorine. The effluents from the cement reactor treated with chlorine and the epoxy reactor treated with monochloramine were also found to be not statistically different from the influent. However, the effluents from the remaining reactors were found to be significantly lower than the influent with differences ranging from 0.7 to 1.0 logs.

In phase three, no statistically significant differences were found between any pairs of reactors containing the same materials that were treated with monochloramine or chlorine (data not shown). In addition, no statistically significant differences were found between reactors containing different materials that were treated with free chlorine. Among reactors containing different materials that were treated with monochloramine, the iron had significantly higher (0.8 logs) effluent counts than the PVC and cement, while all other differences in effluent counts were not statistically different. Statistical comparisons of the reactor effluents to the influent in phase three (0.5 mg/L C plus disinfectant) showed that there was no significant difference between the influent and the effluent from the PVC reactors treated with either monochloramine and chlorine. The remaining reactors treated with chlorine had significantly higher cell counts (0.6 logs) in their effluents than in the influent (3.87 log (cfu/mL). All of the effluent counts from the remaining reactors treated with monochloramine were not statistically different from the influent except for the effluent from the iron reactor, which was higher than the influent by about 1 log.

In phase four (2 mg/L C plus disinfectant) no statistically significant differences were found between the effluents from reactors containing the same materials that were treated with either chlorine or monochloramine. In addition, no statistical differences were reported between the effluents from reactors containing different materials and treated with the same disinfectant, either chlorine or monochloramine. When the effluents from each of the reactors in phase four were compared to the influent (4.15 log (cfu/mL)), no significant statistical differences were found except for the iron and cement reactors that were treated with chlorine (data not shown). The effluents from these reactors were both found to contain higher cell concentrations than the influent by about 0.7 and 0.8 logs, respectively.

Table 4.5
Statistical Comparisons of HPCs in Effluents from Reactors. Lines between materials indicate that no significant statistical differences were found.

a) Phase 1

Material	Epoxy	PVC	Cement	Iron
Ave. log(cfu/ml)	4.37	4.58	4.90	4.99

b) Phase 2

(1) Reactors treated with chlorine.

Material	PVC	Epoxy	Cement	Iron
Ave. log(cfu/ml)	3.37	3.48	3.81	4.23

(2) Reactors treated with monochloramine.

Material	PVC	Cement	Epoxy	Iron
Ave. log(cfu/ml)	3.12	3.51	3.94	4.27

c) Phase 3

(1) Reactors treated with chlorine.

Material	PVC	Epoxy	Iron	Cement
Ave. log(cfu/ml)	4.15	4.45	4.46	4.49

(2) Reactors treated with monochloramine.

Material	PVC	Cement	Epoxy	Iron
Ave. log(cfu/ml)	4.04	4.07	4.26	4.83

d) Phase 4

(1) Reactors treated with chlorine.

Material	PVC	Epoxy	Iron	Cement
Ave. log(cfu/ml)	4.49	4.50	4.81	4.99

(2) Reactors treated with monochloramine.

Material	PVC	Cement	Epoxy	Iron
Ave. log(cfu/ml)	4.50	4.68	4.70	4.70

Table 4.6 illustrates the statistical relationships between the phases for reactors treated with the same disinfectants. For all materials with either monochloramine or chlorine as a disinfectant, statistical analyses showed that there was a significant difference between phase one and phase two, where phase one was greater than phase two. In most cases these analyses also showed that there was a significant difference between phases two and three, where phase three was greater than phase two. The exceptions were the iron reactors and the epoxy reactor treated with monochloramine for which phase two was not significantly different from phase three. On the other hand, the analyses showed that there was no significant difference in the effluent HPCs between phase one and phase four. In addition, in most cases, there was no significant difference between phases three and four. The exception was the cement reactors where phase four was greater than phase three.

Influent/Effluent Total Direct Counts

The influent and effluent TDC data for phases two through four are illustrated in Figures 4.5 through 4.7. A statistical analysis of the total direct counts (TDCs) in the influent showed that only the counts in phases two and three were comparable. This analysis also showed that the influent counts in phase four were significantly greater than those in both phase two and phase three by about 0.3 and 0.4 logs, respectively. In phase two, no significant differences were found between the effluents from the iron reactors or the cement reactors when treated with either monochloramine or chlorine. However, the epoxy and PVC reactors treated with monochloramine had statistically significant higher effluent counts than their pairs treated with free chlorine by about 0.4 to 0.5 logs.

Table 4.6

Statistical Comparisons of Effluent HPCs Between Phases. Lines between materials indicate that no statistically significant differences were found.

(a) Epoxy reactor treated with chlorine.

Phase	2	1	3	4
Ave. log(cfu/ml)	3.48	4.37	4.45	4.50

(b) Epoxy reactor treated with monochloramine.

Phase	2	3	1	4
Ave. log(cfu/ml)	3.94	4.26	4.37	4.70

(c) Iron reactor treated with chlorine.

Phase	2	3	4	1
Ave. log(cfu/ml)	4.23	4.46	4.81	4.99

(d) Iron reactor treated with monochloramine.

Phase	2	4	3	1
Ave. log(cfu/ml)	4.27	4.70	4.83	4.99

(continued)

Table 4.6 (continued)

(e) Cement reactor treated with chlorine.

Phase	2	3	1	4
Ave. log(cfu/ml)	3.81	4.49	4.90	4.99

(f) Cement reactor treated with monochloramine.

Phase	2	3	4	1
Ave. log(cfu/ml)	3.51	4.07	4.68	4.90

(g) PVC reactor treated with chlorine.

Phase	2	3	4	1
Ave. log(cfu/ml)	3.37	4.15	4.49	4.58

(h) PVC reactor treated with monochloramine.

Phase	2	3	4	1
Ave. log(cfu/ml)	3.12	4.05	4.50	4.58

Results of statistical analyses comparing reactors of different materials that were treated with the same disinfectant for phases 2 through 4 are shown in Table 4.7. In phase 2, no significant differences were found between any of the reactors treated with monochloramine. These results show that among the reactors treated with chlorine there were no significant differences between the effluent counts from the PVC, cement, and epoxy reactors, but that all three were significantly lower than the iron reactor.

Statistical comparisons of the effluents to the influent (5.04 log (cfu/mL)) in phase two showed that only the epoxy reactor treated with monochloramine (5.46 log (cfu/mL)) and the iron reactors treated with either chlorine (5.45 log (cfu/mL) or monochloramine (5.55 log (cfu/mL)) were significantly different from the effluent. The mean effluent counts from all three of these reactors were greater than the mean influent count. In phase three, no significant differences between the effluents from the epoxy reactors or the cement reactors when treated with either monochloramine or chlorine were found. However, these analyses showed that the iron and PVC reactors treated with monochloramine had significantly higher effluent counts than their pairs treated with chlorine. The only significant difference in the effluent counts from reactors treated with free chlorine was between the cement and PVC reactors. In this case the cement reactor had a higher mean effluent count than the PVC reactor. Among the reactors treated with monochloramine, the only significant differences were between the iron and PVC reactors and the cement and iron reactors. Here, the iron reactor had the highest effluent. Statistical comparisons of the effluent counts to the influent count (4.91 log (cfu/mL)) showed that all of the effluents were significantly higher than the influent by about 0.6 to 1.2 logs.

In phase four, no statistically significant differences were found between the effluents from reactors containing the same materials that were treated with either chlorine or monochloramine. In addition, no statistical differences were reported between the effluents from reactors containing different materials and treated with the same disinfectant (chlorine or monochloramine). Statistical analyses also showed that the effluent counts from all of the reactors in phase four were significantly higher than the influent (5.28 log (cfu/mL)) by about 0.5 to 0.8 logs.

Table 4.7
Statistical Comparisons of TDCs in Effluents from Reactors for Phase 2 through 4. Lines between materials indicate that no significant statistical differences were found.

(1) Phase 2

(a) Reactors treated with chlorine.

Material	PVC	Epoxy	Cement	Iron
Ave. log(cfu/ml)	4.93	4.99	5.08	5.45

(b) Reactors treated with monochloramine.

Material	Cement	PVC	Epoxy	Iron
Ave. log(cfu/ml)	5.37	5.37	5.46	5.55

(2) Phase 3

(a) Reactors treated with chlorine.

Material	PVC	Iron	Epoxy	Cement
Ave. log(cfu/ml)	5.30	5.56	5.59	5.63

(2) Reactors treated with monochloramine.

Material	Cement	PVC	Epoxy	Iron
Ave.log(cfu/ml)	5.57	5.59	5.77	6.06

(3) Phase 4

(a) Reactors treated with chlorine.

Material	PVC	Epoxy	Iron	Cement
Ave. log(cfu/ml)	5.80	5.82	5.92	6.01

(b) Reactors treated with monochloramine.

Material	PVC	Epoxy	Cement	Iron
Ave. log(cfu/ml)	5.90	5.93	5.92	6.11

Table 4.8 illustrates the statistical relationships between phases of reactors treated with the same disinfectants. The analyses show that for the cement reactors and PVC reactors treated with chlorine, there were significant differences between the effluents in all three phases. The trends in the data for all of the combinations of materials and disinfectants show that, in general, the order of increasing effluent TDCs is as follows: phase two is less than or equal to phase three, which was less than or equal to phase four. For all materials treated with either disinfectant the effluent concentrations in phase four were significantly different from those in phase two.

Biofilm Heterotrophic Plate Counts

The HPC biofilm data for phases one through four are illustrated in Figures 4.8 through 4.11. The results of statistical comparisons of reactors with different materials are shown in Table 4.9. Statistical analyses of the biofilm HPCs for phase one showed that there were no significant differences were found between pairs of reactors containing the same materials. The cement and iron were not significantly different from each other, nor were the PVC and cement, or the epoxy and PVC. Of the remaining comparisons, the iron was greater than the PVC and the epoxy, and the cement was greater than the epoxy. In phase two, no statistically significant differences were found between iron or cement reactors that were treated with the same disinfectant. However, the biofilm densities in the epoxy and PVC reactors treated with monochloramine were greater than biofilm densities in reactors with the same materials that were treated with chlorine. Statistical analyses of the biofilms in reactors with different materials that were treated with the same disinfectant showed that all of the reactors were significantly different from each other. The highest counts were always found in the iron reactors while the lowest counts were always found in the PVC reactors.

Table 4.8

Statistical Comparisons of Effluent TDCs Between Phases. Lines between materials indicate that no statistically significant differences were found.

(a) Epoxy reactor treated with chlorine.

Phase	2	3	4
Ave. log(cfu/ml)	4.99	5.59	5.82

(b) Epoxy reactor treated with monochloramine.

Phase	2	3	4
Ave. log(cfu/ml)	5.46	5.77	5.93

(c) Iron reactor treated with chlorine.

Phase	2	3	4
Ave. log(cfu/ml)	5.45	5.56	5.92

(d) Iron reactor treated with monochloramine.

Phase	2	3	4
Ave. log(cfu/ml)	5.55	6.06	6.12

(e) Cement reactor treated with chlorine.

Phase	2	3	4
Ave. log(cfu/ml)	5.08	5.53	6.01

(f) Cement reactor treated with monochloramine.

Phase	2	3	4
Ave. log(cfu/ml)	5.37	5.57	5.92

(g) PVC reactor treated with chlorine.

Phase	2	3	4
Ave. log(cfu/ml)	4.93	5.30	5.80

(h) PVC reactor treated with monochloramine.

Phase	2	3	4
Ave. log(cfu/ml)	5.37	5.59	5.90

Table 4.9
Statistical Comparisons of Biofilm HPCs for Phase 1 through 4. Lines between materials indicate that no statistically significant differences were found.

(1) Phase 1

Material	Epoxy	PVC	Cement	Iron
Ave. log(cfu/cm^2)	5.14	5.26	5.58	5.90

(2) Phase 2

(a) Reactors treated with chlorine.

Material	PVC	Epoxy	Cement	Iron
Ave. log(cfu/cm^2)	2.84	3.66	4.86	5.97

(b) Reactors treated with monochloramine.

Material	PVC	Cement	Epoxy	Iron
Ave. log(cfu/cm^2)	3.76	4.46	5.11	6.14

(3) Phase 3

(a) Reactors treated with chlorine.

Material	PVC	Epoxy	Cement	Iron
Ave. log(cfu/cm^2)	4.17	4.52	4.79	5.76

(b) Reactors treated with monochloramine.

Material	PVC	Cement	Epoxy	Iron
Ave. log(cfu/cm^2)	4.12	4.49	5.12	5.40

(4) Phase 4

(a) Reactors treated with chlorine.

Material	PVC	Epoxy	Cement	Iron
Ave. log(cfu/cm^2)	4.88	5.15	5.42	6.65

(b) Reactors treated with monochloramine.

Material	PVC	Epoxy	Cement	Iron
Ave. log(cfu/cm^2)	5.14	5.14	5.33	6.45

In phase three, statistical analyses of pairs of reactors of the same materials that were treated with different disinfectants showed no significant differences between any of the four pairs. Results show that the iron reactors were significantly different from the epoxy, PVC, and cement reactors treated with the same disinfectants. The biofilm densities in these iron reactors were higher than in any of the other reactors treated with the same disinfectants by 1 to 2.3 logs. The biofilm densities in the PVC and epoxy reactors were also found to be significantly different when treated with monochloramine. All remaining differences for either disinfectant were found to be not statistically significant. The results of the statistical analyses of the biofilms in phase four were very similar to those from phase three. The average counts ranged from 4.9 to 6.7 log (cfu/cm^2). Again, there were no significant differences between any reactor pairs containing the same materials that were treated with different disinfectants. In addition, the biofilm densities in the iron reactors were found to be significantly higher than those in other reactors treated with the same disinfectants. Also in phase four, the PVC, epoxy, and cement reactors were not significantly different from each other under treatment with either disinfectant.

Table 4.10 shows the statistical relationships between phases of reactors treated with the same disinfectants. The statistical analyses show that for reactors with all materials except for iron, the biofilm densities in phase four were not significantly different from those in phase one when treated with either disinfectant. The iron reactors treated with either disinfectant had higher biofilm densities in phase four than in phase one. All reactors that were treated with monochloramine had no significant differences between biofilm densities in phases two and three. For all material and disinfectant combinations the biofilm densities in phase two were the lowest or not significantly different from the phase with the lowest densities. In addition, phase four had the highest counts or was not significantly different from the phases with the highest counts

DISCUSSION

Two methods of measurement were used to examine the cellular densities in the reactor influent and effluents. In general, the densities measured using total direct counts (TDCs) were 10 to 100 times higher than those measured using the heterotrophic plate count (HPC) method. This indicates that not all of the cells in these samples could grow under the conditions used in the HPC method. This can be attributed to cell injury, variations in the growth phase at the time of capture, and genotypic or phenotypic characteristics of some of the bacterial species that prevented them from growing under these conditions.

The two hour detention time in the annular reactors was sufficient to allow biofilms to accumulate and grow, but not to allow significant planktonic growth. Thus, any difference between the influent and the effluent cell counts can be attributed to detachment from the biofilm. This is important to remember when interpreting the influent and effluent data.

Table 4.10
Statistical Comparisons of Biofilm HPCs Between Phases. Lines between materials indicate that no statistically significant differences were found.

(a) Epoxy reactor treated with chlorine.

Phase	2	3	4	1
Ave. log(cfu/cm^2)	3.66	4.63	5.15	5.14

(b) Epoxy reactor treated with monochloramine.

Phase	2	3	1	4
Ave. log(cfu/cm^2)	5.11	5.12	5.14	5.14

(c) Iron reactor treated with chlorine.

Phase	3	1	2	4
Ave. log(cfu/cm^2)	5.76	5.90	5.97	6.65

(d) Iron reactor treated with monochloramine.

Phase	1	2	3	4
Ave. log(cfu/cm^2)	5.90	6.14	6.40	6.45

(e) Cement reactor treated with chlorine.

Phase	3	2	4	1
Ave. log(cfu/cm^2)	4.79	4.86	5.42	5.58

(f) Cement reactor treated with monochloramine.

Phase	2	3	4	1
Ave. log(cfu/cm^2)	4.46	4.49	5.33	5.58

(g) PVC reactor treated with chlorine.

Phase	2	3	4	1
Ave. log(cfu/cm^2)	2.84	4.17	4.88	5.26

(h) PVC reactor treated with monochloramine.

Phase	2	3	4	1
Ave. log(cfu/cm^2)	3.76	4.12	5.14	5.26

Comparisons of the TDC and HPC influent and effluent data revealed some other significant discrepancies. One observation is that the TDC influents were generally significantly lower than the TDC effluents from the reactors in phases two through four. However, the HPC influents were much closer to the HPC effluents and were often not statistically different in phases two through four. This was probably because some of the cells were inactivated due to disinfection and were unable to grow, although they were still visible using the direct count method. This biofilm contribution to the effluent counts was masked in the HPC data by the disinfection effect, which led to artificially low counts. The result was that significant differences were found between the TDC and HPC data sets in any analyses that made comparisons to or between the influents.

Another difference between the TDC and HPC planktonic data was revealed when effluents from reactors with the same materials that were treated with different disinfectants for each phase were compared. No statistically significant differences were found in the HPC data set. However, in the TDC data set some significant differences were found, but there was no regularity to these differences and neither disinfectant could be conclusively shown to have a greater efficacy.

There were also a few differences between the TDC results and the HPC results when effluents from reactors with different materials that were treated with the same disinfectants were compared. However, these differences did not lead to any major differences in the overall conclusions for the influent and effluent data.

Despite the many differences between the TDC and HPC influent and effluent data, there are some general trends that appear in both data sets. When the influent counts were compared to the effluent counts, both data sets demonstrated that there was very little difference between the efficacies of monochloramine and chlorine against the planktonic cells in reactors containing any of the studied materials. In all three phases in which disinfectants were added, very few statistically significant differences were found between reactors containing the same materials when treated with monochloramine or chlorine. The few statistically significant differences that were found did not form a discernible pattern throughout the phases and therefore did not indicate that one disinfectant had a greater efficacy than the other.

As stated above, statistically there was very little difference in the efficacy of either disinfectant as a function of material on effluent cell counts, regardless of materials or disinfectant. In fact, in the HPC data in phase four, at the highest carbon level, no significant statistical differences in effluent cell counts were found between reactors of any material when treated with either chlorine or monochloramine at a residual of 0.2 mg/L. These statistical results are based on probabilities and tend to be conservative; therefore, they may not show the trends. However, inspection of the means show trends in the effluent data as a function of material. Iron reactors generally had the highest effluent densities or the results were not significantly different from reactors with the highest effluent densities. In addition, PVC always had the lowest densities or these values were not significantly different from the reactors with the lowest densities. The effluents from the cement and epoxy reactors fell in between the iron and PVC with no regular pattern with respect to each other.

For all materials with either monochloramine or chlorine as a disinfectant, statistical analyses of the HPC effluent data showed that there was a significant difference between phase one and phase two. This demonstrates the efficacies of both chlorine and monochloramine against detached biofilms cells. Both the HPC and the TDC data showed that the effluent counts in phase three, when carbon was added in the presence of disinfectants, were always greater than or equal to the effluent counts in phase two. In addition, it was found that in both of these data sets the effluent counts in phase four, where carbon levels were higher, were always greater than or equal to the effluent counts in phase three. This shows a general trend that increases in carbon levels are accompanied by higher effluent cell concentrations. This increase can be attributed to detachment from the biofilms in the reactors since the residence time in the reactors was too short to allow significant planktonic growth to occur. At steady state the rate of biofilm growth is equal to the rate of detachment from the biofilm. Thus, chlorine and monochloramine at a residual of 0.2 mg/L do not completely control biofilm growth. However, this residual was sufficient at all studied carbon levels to maintain effluent HPC concentrations at or below the levels in the control, where no carbon or disinfectants were added.

Biofilm HPC Data

The biofilm HPC data showed no conclusive evidence that one disinfectant had a greater efficacy against biofilms than the other one did. This is contrary to many other studies which have shown that monochloramine is a better disinfectant for controlling biofilms. The discrepancy found between this study and earlier disinfection studies is most likely related to the dose used in this experiment. In distribution systems, the target minimum disinfectant residual is often 0.2 mg/L, but it can range up to 4 mg/L. In fact, the actual residual in the distribution system is often much higher than 0.2 mg/L so that this minimum residual can be achieved much further downstream. LeChevallier et al. (1990a) showed that there was a minimum threshold level at which monochloramine was effective for controlling biofilms in iron pipes. This threshold was determined to be 2.0 mg/L in that study, but is likely to vary with water quality and pipe characteristics. The same study found that chlorine was ineffective at doses of up to 4 mg/L. Another study (Sanderson et. al, 1997) found that *Pseudomonas aeruginosa*, a common constituent of drinking water biofilms, showed evidence of adaptation to monochloramine concentrations below 0.5 mg/L. Both of these concentrations are well above the concentration applied in this experiment and support the theory that the apparent equality of the disinfectants against the biofilms in this experiment is related to concentration.

The low residual used in this experiment represents the worst case scenario for drinking water distribution systems where the residual is near the minimum of 0.2 mg/L. This may occur near the end of the distribution system or in longer sections of the distribution where no disinfectant supplements are supplied and residuals have degraded. These results are also relevant to dead ends or low flow loops in the distribution system where residuals have degraded.

The biofilms in the iron reactor treated with monochloramine had a significant increase in density between phases one and two. Since the only difference between these two phases was the addition of a 0.2 mg/L monochloramine residual, this increase can only be attributed to the presence of the monochloramine. This corresponds to previous work that found that the disinfectant efficacy of monochloramine is further reduced against biofilms in the presence of iron (LeChevallier et al., 1993). This effect was believed to have been caused largely by the reaction of the disinfectants with the corrosion products, which prevents the biocide from penetrating the biofilm layer. This same study showed that this phenomenon also exists for iron in the presence of chlorine to an even greater degree than the monochloramine. However, in these experiments, there was no conclusive evidence to support or deny this phenomenon for iron. Although it appeared that there was a slight increase in the biofilm densities in the iron reactor when chlorine was added, statistical analyses did not show that this difference was significant.

Evidence from this study supports the phenomenon of the increased bioavailability of humic substances by adsorption to iron surfaces. The biofilms in the iron reactors were found to be significantly denser than the biofilms in reactors containing any of the other materials when the reactors were supplemented with humics at 0.5 and 2 mg/L C. In phase one the biofilm densities in the iron reactors were not significantly different from the densities in the cement reactors and there was very little difference between the densities on all four materials. However, in phases three and four where the supplementary humic substances were added, the densities in the iron reactors were significantly greater than in any other. Meanwhile there was again little difference between the densities on the other materials in these phases. This evidence suggests that the increased bioavailability of humic substances in the presence of iron was a major factor in the significantly greater biofilm growth on the iron coupons in phases three and four. As observed in phase two, the presence of monochloramine and chlorine in phases three and four probably contributed to the higher iron densities as well.

Of the materials other than iron there appeared to be very little difference in the impact of either disinfectant on the biofilm densities as a function of material. Based solely on the statistical results, no regular pattern could be discerned in the biofilm data regarding which reactors had the highest or the lowest densities in any of the phases. However, the trends in the data showed that the PVC always had the lowest densities or did not differ significantly from the material with the lowest densities. The iron reactors had the highest biofilm densities in all phases. The cement and epoxy fell between the PVC and iron with no apparent order of ascendance. These results support previous studies which have shown that biofilms on iron surfaces are more problematic than those on other types of surfaces. LeChevallier et. al (1996) showed that distribution systems with more miles of unlined cast iron pipe had much higher coliform occurrences than systems with significantly less unlined cast iron pipe. Another study showed that the biofilm densities on iron were significantly higher than those on plastic-based materials such as PVC (Niquette et. al, 2000). That study also showed that the biofilm densities on cement were intermediate to plastic-based materials and iron.

The biofilms in the iron reactors in phase four were significantly greater than the biofilm in the same reactors in the control phase. This indicates that the disinfectant level of 0.2 mg/L was insufficient to control the increased growth at the highest carbon level in the iron reactors when compared to the control phase. However, for the biofilms on the remaining materials, the biofilms in phase four did not significantly differ from the biofilms on the same materials in phase one. This indicates that the disinfectant level of 0.2 mg/L was sufficient to control the increased growth at the highest carbon level in epoxy, cement, and PVC reactors when compared to the control phase.

Comparisons of HPC Biofilm and HPC Effluent Data

A comparison of the HPC biofilm values to the HPC effluent trends showed very similar results between both data sets. The results from the statistical analyses for both data sets were often similar and the data reflected the same trends. For both the effluent and biofilm HPCs the iron reactors had the highest densities while the PVC reactors had the lowest densities. The biofilm and effluent densities in the cement and epoxy reactors were intermediate to the densities in the iron and PVC reactors with no defined order of ascendance. In addition, in both data sets there was a general trend of increasing densities as the carbon levels increased.

When the biofilm and effluent densities were compared to the carbon depletion in the reactors only one significant trend appeared, with depletion in the iron reactors consistently high compared to the other reactors. This most likely corresponds to the significantly greater biofilm growth resulting from higher carbon utilization in the iron reactors compared to the other types of reactors. In addition, it has been shown that humic substances readily adsorb to iron oxide surfaces (Gu et al., 1994). It is likely that this phenomenon also contributed, in part, to the greater depletion of carbon in the iron reactors.

SUMMARY

Results for effluent total direct counts (TDCs), effluent heterotrophic plate counts (HPCs), and biofilm HPCs from the laboratory experiments are summarized below.

- Densities measured using the TDC method are 10 to 100 times higher than those measured with the HPC method. This phenomenon is probably due to cell injury, variations in growth phase, and genotypic or phenotypic variations which prevented growth by the HPC method.

- Effluent TDC levels for phase 2 (disinfectants only) were ≤ phase 3 (disinfectants plus 0.5 mg/L TOC) ≤ phase 4 (disinfectants plus 2.0 mg/L TOC), and phase 2 was < phase 4 in all cases

- Effluent HPC levels for 2 (disinfectants only) were ≤ phase 3 (disinfectants plus 0.5 mg/L TOC) ≤ phase 4 (disinfectants plus 2.0 mg/L TOC) in all cases, and generally, phase 4 was equal to phase 3. Effluent HPC levels for phase 4 = phase 1 in all cases.

- Biofilm HPC levels for phase 4 = phase 1 for all materials except iron. For biofilm HPC levels for reactors containing iron, phase 4 > phase 1.

- In all cases, biofilm densities in phase 2 (disinfectants only) were the lowest or not significantly different from the lowest.

- In all cases, biofilm densities in phase 4 (disinfectants plus 2.0 mg/L TOC) were the highest or not significantly different from the highest.

- On the materials other than iron, there was no regular pattern regarding which reactors had the highest or lowest biofilm densities.

- The biofilm HPC densities in the PVC reactors were always the lowest or not significantly different from the reactors with the lowest densities.

- The biofilm HPC densities in the epoxy and cement reactors were often not significantly different from each other and had no significant pattern of ascendance.

- The iron reactors had the highest biofilm densities or were not significantly different from the highest densities in all phases. These results support previous work which has shown that the biofilm densities on iron are significantly greater than those on plastic-based materials such as PVC. This earlier work also showed that the densities on cement are intermediate to those on PVC and iron.

- Comparisons of reactors with the same materials that were treated with different disinfectants showed that in general there was no significant difference in the effluent or biofilm densities between the reactors.

- Among cases where significant differences were found between disinfectant treatments for a given material, there was no pattern to show that one disinfectant had greater efficacy than the other.

- Biofilms from the iron reactors treated with monochloramine in phase 2 (disinfectants only) were statistically significantly denser than those in phase 1 (control). Biofilms from the iron reactors treated with chlorine in phase 2 appeared to be denser than those in phase 1, but the difference was not found to be statistically significant. Biofilms in iron reactors treated with either disinfectant in phase 2 were significantly denser than biofilms on any other materials treated with the same disinfectant. In phase 1 this difference was much smaller. This evidence supports previous findings which showed that the efficacies of monochloramine and chlorine are further reduced against biofilms in the presence of iron.

- Biofilms in iron reactors which received supplementary humics at both 0.5 and 2 mg/L were significantly denser than those on any other materials under the same conditions. There was very little difference between the densities on the other materials. This evidence supports previous work which has shown that there is an increase in the bioavailability of humic substances in the presence of iron. Since the disinfectants were still being added with the supplementary carbon, the chlorine and monochloramine effects were probably also still a factor.

CONCLUSIONS

- There was no significant difference in the efficacies of chlorine and monochloramine against planktonic cells or biofilms at a residual of 0.2 mg/L.

- Increases in carbon levels led to general increases in biofilm and planktonic densities. This effect was most pronounced for biofilms in reactors containing iron.

- Of the epoxy, PVC, and cement reactors there was no definite order of ascendance in regard to biofilm or planktonic growth. However, the PVC was always the lowest or not significantly different from the lowest.

- In the presence of disinfectants and supplementary organic carbon, the iron reactors had the highest effluent counts of any of the materials.

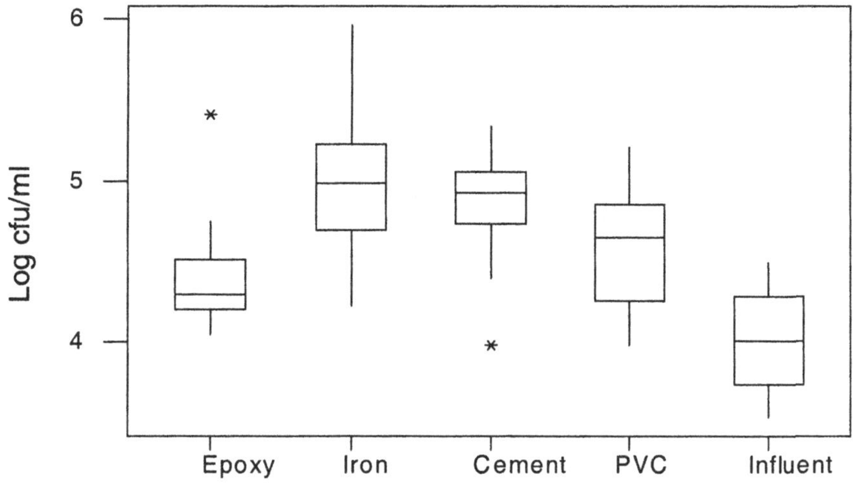

Figure 4.1 HPC influent and effluent densities in Phase 1 (No disinfectant added, no TOC added)

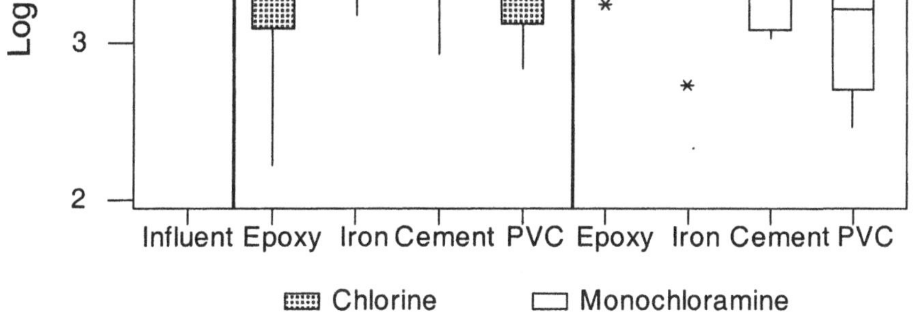

Figure 4.2 HPC influent and effluent densities in Phase 2 (.2 mg/L disinfectant, no TOC added)

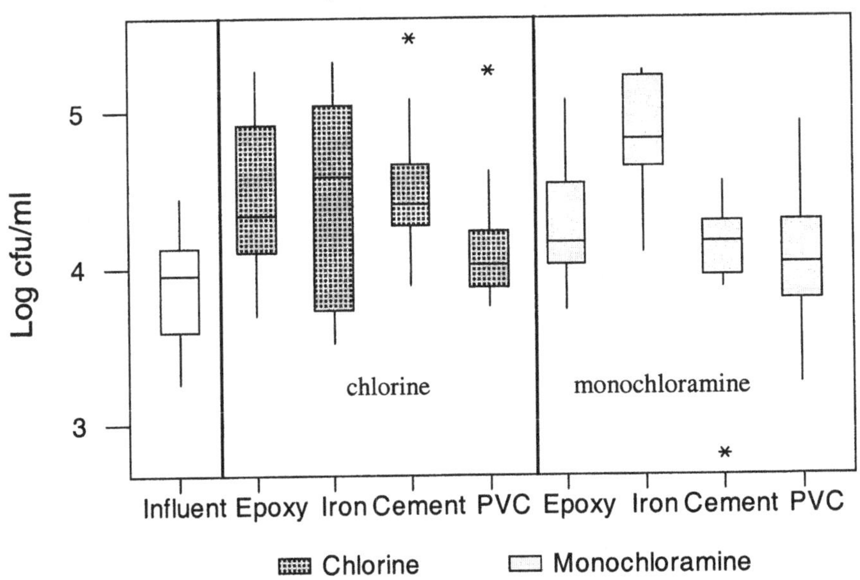

Figure 4.3 HPC influent and effluent densities in Phase 3 (.2 mg/L disinfectant, .5 mg/L TOC)

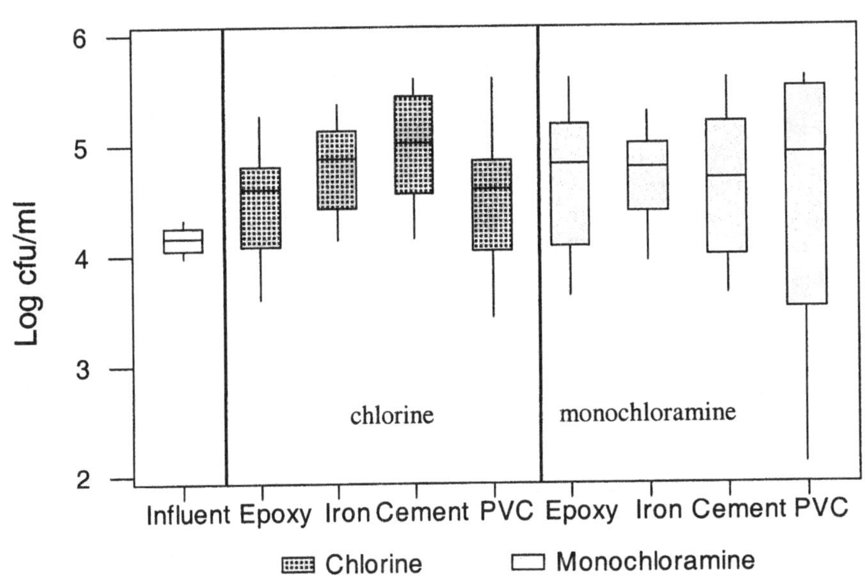

Figure 4.4 HPC influent and effluent densities in Phase 4 (.2 mg/L disinfectant, 2 mg/L TOC)

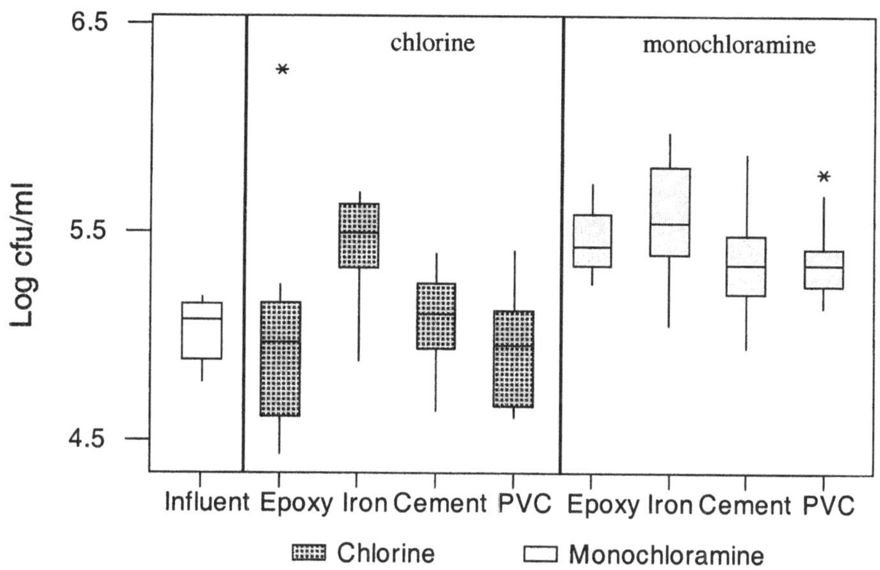

Figure 4.5 Total direct counts influent and effluent densities in Phase 2 (.2 mg/L disinfectant, no TOC)

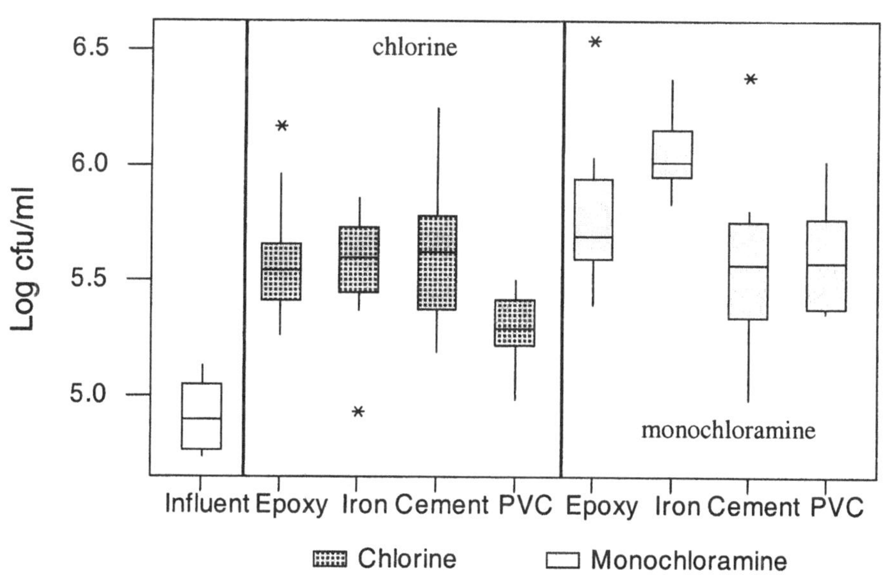

Figure 4.6 Total direct counts influent and effluent densities in Phase 3 (.2 mg/L disinfectant, .5 mg/L TOC)

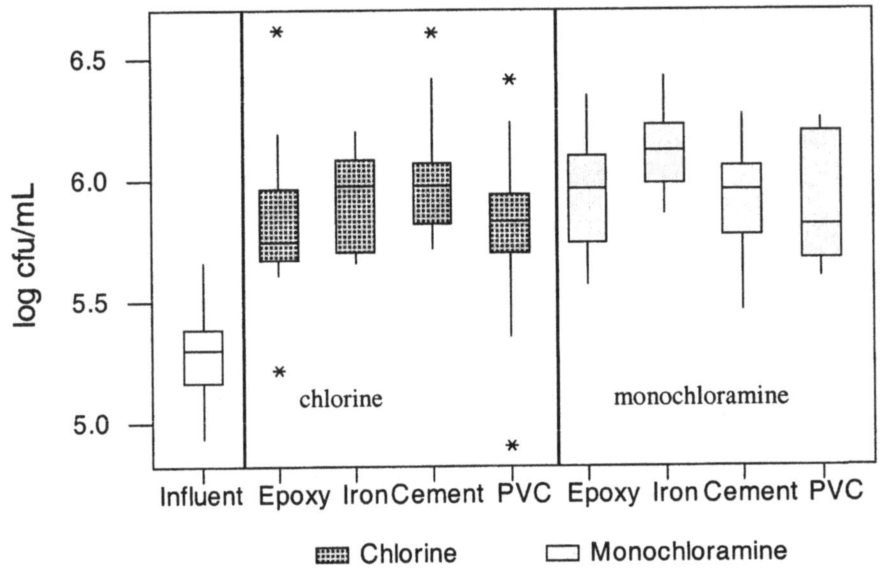

Figure 4.7 Total direct counts influent and effluent densities in Phase 4 (.2 mg/L disinfectant, 2.0 mg/L TOC)

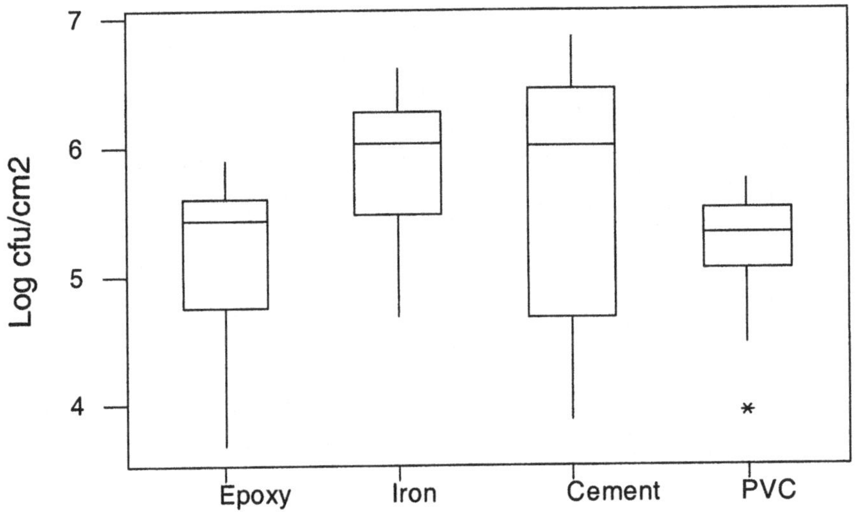

Figure 4.8 HPC biofilm densities in Phase 1 (No disinfectants added, no TOC added)

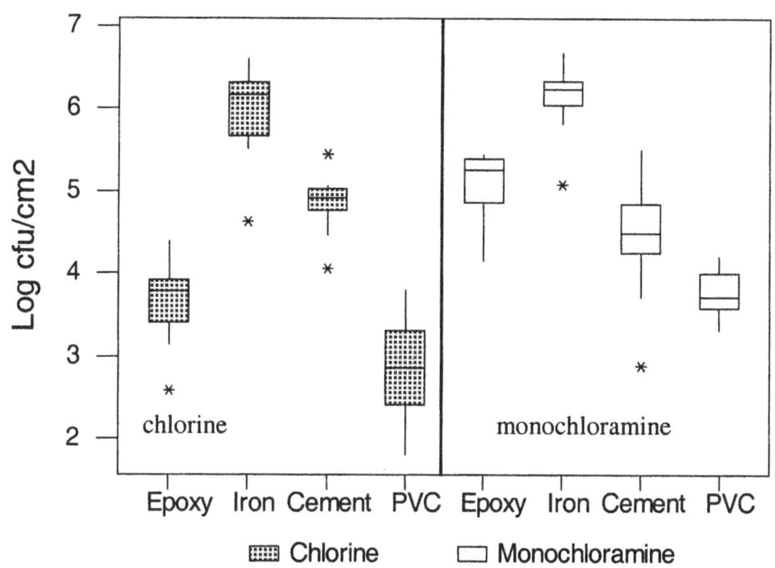

Figure 4.9 HPC biofilm densities in Phase 2 (.2 mg/L disinfectant, no TOC added)

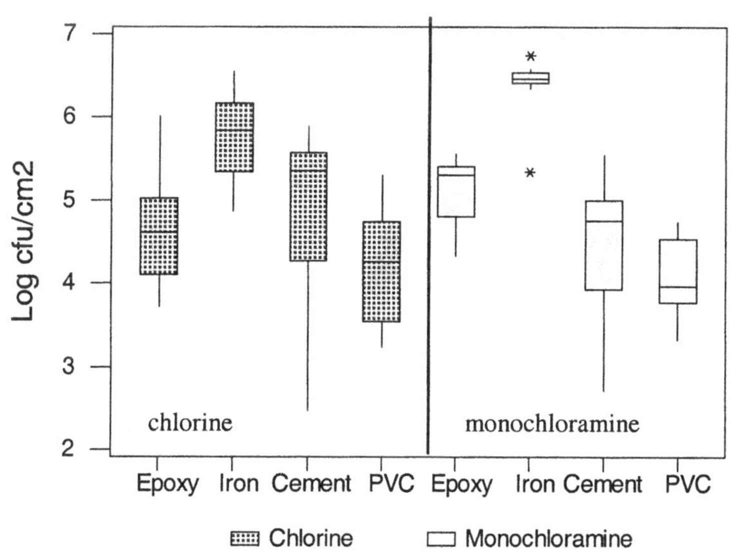

Figure 4.10 HPC biofilm densities in Phase 3 (.2 mg/L disinfectant, .5 mg/L TOC)

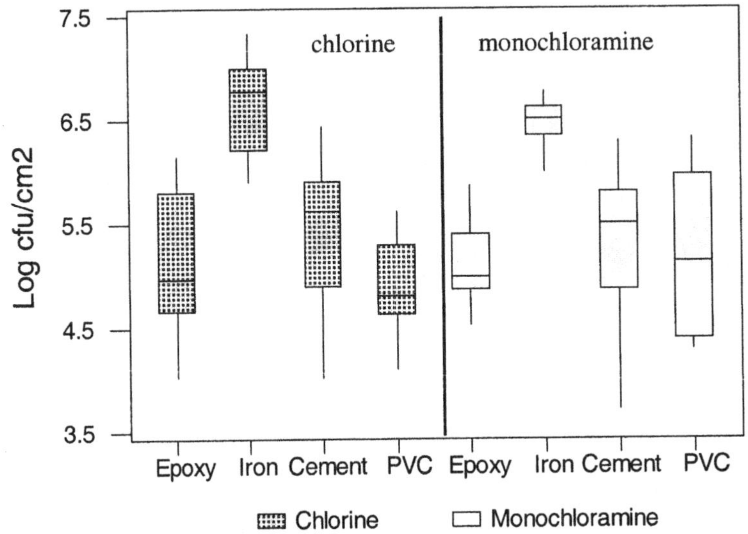

Figure 4.11 HPC biofilm densities in Phase 4 (.2 mg/L disinfectant, 2.0 mg/L TOC)

CHAPTER 5
FIELD STUDIES

INTRODUCTION

Field studies incorporating annular reactors with three different materials (ductile iron, cement lining, and epoxy) were conducted concurrently with the laboratory studies described in the previous chapter. The field studies took place at four utility locations: Cambridge, Massachusetts; Cincinnati, Ohio; Greenville, South Carolina; and Tampa Bay, Florida. The objective of these field studies was to determine the numbers of organisms found on coupons made of various materials in order to evaluate how water quality and pipe material interact to foster biofilm growth at each specific utility location. In addition to the annular reactor field studies, two utilities (Greenville and Cambridge), excavated pipes from the distribution system and bacterial numbers were measured to compare to the bacterial number measured on the annular reactor coupons. These results can be found in Appendix D. Descriptions of the methods and materials for each of the utility annular reactor field studies are contained below, followed by the results of these studies.

METHODS AND MATERIALS

Experimental Setup at Each Utility

Cambridge Water Department

The Cambridge Water Department (CWD) in Cambridge, Massachusetts, serves water to approximately 100,000 residents. At the beginning of this study, CWD purchased water from the Massachusetts Water Resources Authority (MWRA), which is treated with chlorine as a primary disinfectant and chloramines as the secondary disinfectant in the distribution system. During the course of this AWWARF study, CWD completed construction of a new water treatment plant composed of pre-ozonation, coagulation, dissolved air flotation, ozonation, and deep bed biologically active GAC filtration that includes biologically activated filtration (BAF). This plant came on-line in March, 2001. The system uses chloramines for disinfection and adjusts pH to 9.0 - 9.3 with KOH for corrosion control.

The main objective of the CWD field study was to evaluate and compare the impacts of unfiltered water and biologically treated water on regrowth on the different piping materials. Six annular reactors were installed at an existing aboveground CWD valve station in February, 2000 (see Figure 5.1). A 16" pipe, located at the "start" of the distribution system, was used to supply water to the annular reactor. Water from this pipe represented the system start point for both water sources (MWRA and CWD). During the interim period before the new treatment facility was on-line, water was fed from the MWRA supply interconnection to the pipe before branching to the distribution system. When the system converted to CWD source water, water flowed from the plant to the water storage tanks and then through 1000 feet of pipe to the tap location before being distributed. Two reactors were allocated for each of the materials selected (ductile, cement lining, and epoxy), with each reactor housing 20 coupons. Polyethylene sheeting was used to cover the annular reactors to prevent sun-induced biogrowth. The building temperature was maintained at a temperature similar to ground temperature, i.e. 45 ^{0}F.

Water quality samples were collected from the influent to the reactors for pH, temperature, alkalinity, heterotrophic plate counts, chlorine residual, and chlorine residual every two weeks. Coupons were pulled from the reactors approximately once every two weeks. Collection of water quality data and analysis of reactor coupons began in March 2000 and was completed in June, 2001, allowing for approximately twelve months of operations using MWRA water and four months of operation using CWD water.

Cincinnati Water Works

The Cincinnati Water Works (CWW) serves water to approximately one million people in the greater Cincinnati, Ohio area. Water originates from two water treatment facilities: the Bolton plant which treats groundwater and the Miller plant which treats water from the Ohio River. The Miller plant includes conventional filtration followed by treatment with granular activated carbon (GAC). A large fraction of the total organic carbon is removed by the GAC contactors.

The testing program for CWW had two main objectives; first, to test water from the Miller plant to determine the impact TOC reductions may have on regrowth for different piping materials; and second, to install annular reactors at a remote location in the distribution system to determine the impact of water age on regrowth. To achieve these objectives, three annular reactors were installed at the Miller plant and three annular reactors were installed at the Sayler Park Fire Station, a location in the distribution system where the water age was approximately three days (see Figure 5.2). Two reactors were allocated for each of the materials selected (ductile iron, cement lining, and epoxy) so that each testing location contained reactors with each of the materials.

Water quality samples were collected from the influent to the reactors for pH, temperature, chlorine residual, and HPC every two to three days. Total organic carbon levels were collected monthly, and coupons were pulled from the reactors every one to two weeks. Operation of the annular reactors began in February 2000 and continued through May 2001.

Greenville Water System

The Greenville Water System (GWS) serves a population of 320,000, which includes 12 municipalities including the city of Greenville, South Carolina. The average daily demand of the water system is 55 MGD. At the beginning of this study, the GWS had two sources of water. One source was unfiltered (chlorine, fluoride, a corrosion-controlling chemical and sodium hydroxide) and one was treated with conventional filtration. A new water treatment facility composed of pre-chlorination, coagulation, dissolved air flotation, and monomedia anthracite filtration was constructed in February 2000. The objective of the GWS study was to compare the relative impacts of filtered versus unfiltered water on regrowth on the piping materials.

A total of six parallel annular reactors were installed with two reactors allocated for each of the materials selected (ductile iron, cement lining, and epoxy). The reactors were installed at an abandoned chlorine booster station near the GWS Central Laboratory located in Travelers Rest (see Figure 5.3). A transmission main located at the "start" of the distribution system downstream of where the two sources are combined (before DAF) and downstream of the new treatment facility (after DAF) were used to supply water to the annular reactor. Reactors began operation on February 7, 2000. Prior to July 4, 2000, the water delivered to the annular reactors was unfiltered, and after that date, all of the water in the system was fed through the new dissolved air floatation (DAF) filtration plant.

Water quality samples were collected from the influent to the reactors for pH, temperature, chlorine residual and HPC every two to three days. Samples for alkalinity were collected weekly and TOC samples were collected monthly. Coupons were pulled from the reactors every two weeks. Operation of the reactors began in February 2000 and continued through June 2001.

Tampa Bay Water

Tampa Bay Water is a wholesale water supply utility that provides water to Hillsborough, Pasco and Pinellas counties as well as the cities of New Port Richey, St. Petersburg and Tampa, Florida. The source water is a highly organic groundwater, with TOC levels ranging from 3-5 mg/L, with typical values of 4-4.5 mg/L.

The objective of the Tampa Bay field study was to evaluate the impact of a relatively high TOC groundwater on regrowth on the various piping materials. A total of three parallel annular reactors were installed, one of each type of material (ductile iron, cement-lined, and epoxy), at Pinellas County's Keller Water Treatment Plant, located in the Eldridge Wilde wellfield (see Figure 5.4). The treatment scheme at the Keller plant consists of aeration and chlorination.

Water quality samples were collected from the influent to the reactors twice a week for pH, chlorine residual, alkalinity, TOC and HPC. Tampa Bay Water began operation of its three reactors in October, 2000 and continued through May 2001.

Chemical and Biological Analyses

Samples of the influent water to the reactors were collected and analyzed for pH, temperature, chlorine residual, alkalinity, TOC, and HPC by utility personnel at each reactor location, and certified laboratories were used for analyses of these samples. Coupons were removed from the reactors, replaced with a coupon of similar material, and placed into sterile wide mouth screw-topped test tube filled with sterile phosphate-buffered saline (PBS) or sterile water prior to sealing for transport and analysis. The tubes were stored on ice and sent by overnight delivery to Dr. Camper at Montana State University for analysis of biofilm HPC levels. Coupon biofilm analysis was conducted within 24 hours of sampling.

Statistical Analyses

As in the laboratory studies described in the previous chapter, the logarithms of the cell densities (attached HPC levels) were evaluated to determine statistically significant differences both within each utility's field study data, and for data between utilities. Parametric t-tests were completed on all data sets using CSS STATISTICA™ Release 5.1. The t-test is the most commonly used method to evaluate the differences in means between two groups. p-levels were determined, which represent the probability of error associated with rejecting the hypothesis of no difference between the two categories of observations (whether they be data from reactors with different materials at one utility location, data from reactors with the same materials at different utility locations, or data from the same material under different water quality conditions) when in fact, the hypothesis is true. If the resultant t-value was statistically significant, then it was concluded that the means in the two groups were different.

For field study locations where duplicate reactors with the same coupon material were evaluated, the data were evaluated to determine if the two coupons exhibited statistically similar results. If so, the data were grouped and if not, each reactor was evaluated separately when determining statistical differences between the different materials.

As with the laboratory data, the data described in the following sections are displayed graphically using boxplots. These boxplots consist of three major features; a large box, a smaller box inside the larger box, and a pair of vertical lines extending from the upper and lower ends of the larger box. The larger box portion of the plot represents the middle 50% of the observations. The small box represents the median of the data. The lines extending from the box are called whiskers. These whiskers indicate the lowest and highest values in the data set.

RESULTS

Water Quality and Biofilm Heterotrophic Plate Counts for Each Utility Study

Cambridge Water Department

Water quality data for the influent to the six CWD annular reactors during the course of the field study are displayed in Table 5.1 and Figure 5.5. In general, the influent alkalinity and free chlorine was slightly higher and the total chlorine levels were slightly lower in the CWD water when compared to the MWRA water.

Table 5.1

Cambridge Water Quality Data For Influent to Annular Reactors

Parameter	MWRA Water	CWD Water
PH	8.8 (8.2 – 9.2)	9.0 (8.6 – 9.2)
Temperature, deg C	10.5 (0.5 – 19.7)	11.2 (1.2 – 21.5)
Alkalinity, mg/L	28 (25 – 31)	37 (34 – 43)
TOC, mg/L	2.0 (2.0 – 2.1)	-
Free Chlorine, mg/L	0.1 (0 – 0.1)	0.4 (0.3 – 0.5)
Total Chlorine, mg/L	2.1 (1.7 – 2.5)	1.6 (1.4 – 1.9)

At the conclusion of the study (June 25, 2001), approximately 34 sets of bi-weekly slide coupon samples had been collected, 25 sets before the conversion (MWRA water) and 9 sets after the conversion (CWD water). Coupon HPC levels over the course of the study are displayed in Figure 5.6. There were no significant differences in coupon HPC levels for the duplicate cement and iron annular reactors during the phase of the study where MWRA was tested, or for the duplicate cement and epoxy annular reactors during the phase of the project where CWD water was tested. The data from these coupons was grouped for the remainder of the statistical analyses. Data from the duplicate epoxy reactors during the phase of the study where MWRA water was tested and data from the duplicate ductile iron reactors during the phase of the study where CWD water was tested, were evaluated separately.

As seen in Figure 5.6, the iron coupons had higher attached HPC levels than the epoxy or the cement coupons. This is also evident in Figures 5.7 and 5.8 which display comparisons of log attached HPC levels for each material for MWRA and CWD water respectively. Table 5.2 shows the statistical relationship between the log attached HPC levels for these different materials. The attached HPC levels for the cement and epoxy coupons were not statistically different from one another, but the iron coupons were significantly higher than either of these materials for both the MWRA water and the CWD water. There was no difference in the attached HPC levels measured from any of the materials when the reactors were receiving MWRA water vs. CWD water (not shown).

Table 5.2
Statistical Comparisons of Mean Log HPC Levels for Cambridge Water Department. Lines between materials indicate that no statistically significant differences were found.

(a) Cambridge Field Study – Before (MWRA Water)

Material	Cement	Epoxy	Iron
Average log HPC (cfu/cm_2)	2.85	3.3 – 4.3	6.96

(b) Cambridge Field Study – After (Cambridge Water)

Material	Cement	Epoxy	Iron
Average log HPC (cfu/cm_2)	2.2	4.2	7.1 – 7.4

Cincinnati Water Works

The raw influent data to the six reactors are shown in Figures 5.9 and 5.10 and averages are displayed in Table 5.3 below. The Saylor Park Fire Station reactors, which are located in the distribution system, displayed lower chlorine residual levels and higher HPC levels than the reactors located in the Miller Treatment Plant. Alkalinity, pH, and TOC levels were similar in the two locations.

Table 5.3
Cincinnati Water Quality Data for Influent to Annular Reactors

Parameter	Richard Miller Plant		Saylor Park Fire Station	
	Average	Range	Average	Range
PH	8.9	8.5-9.2	8.8	8.3-9.0
Temperature, deg C	16.3	3 – 26	17.5	7-27
Alkalinity, mg/L	71.4	55 – 86	69.9	55-78
HPC, counts/100 mL	5.7	0 – 160	47.3	0-1,200
TOC, mg/L	0.8	0.5 – 1.3	0.8	0.6-1.3
Chlorine, mg/L	1.2	0.4 – 4.3	0.7	0.4-1.2

One coupon was removed from each of the six reactors once every two weeks and analyzed for bacterial growth. By the end of the study period, 34 sets of coupon samples had been collected from the annular reactors. The coupon attached HPC results for the entire testing period for all materials are presented in the Figure 5.11, where the solid lines represent the coupons from the Miller Treatment Plant and the dashed lines represent the coupons from the Saylor Park Fire Station in the distribution system. Average log attached HPC levels from the coupons are displayed in Figure 5.12. On visual examination, the log attached HPC levels from the iron coupons from the Saylor Park Fire Station in the distribution system appear to have higher log attached HPC levels than the rest of the coupons. Table 5.4 displays the statistical comparisons of the different materials at both locations. There was no statistical difference between the materials for the reactors located at the Miller Treatment Plant. For the reactors located in the distribution system at the Saylor Park Fire Station, the cement coupons were statistically similar to both the epoxy and the iron, but the epoxy and iron coupons had statistically different attached HPC levels. The iron coupons located in the distribution system had the highest average log attached HPC levels.

When comparing the results for coupons of the same material located at the Miller Treatment Plant and in the distribution system, there were no differences in the log attached HPC levels on the epoxy coupons, but the iron and cement coupons had statistically different biofilm levels between the two sites, with the distribution system coupons exhibiting higher values.

Table 5.4
Statistical Comparisons of Mean Log HPC Levels for Cincinnati Water Works. Lines between materials indicate that no statistically significant differences were found.

(a) Cincinnati Field Study – Miller WTP

Material	Epoxy	Cement	Iron
Average log HPC (cfu/cm$_2$)	3.2	3.3	2.8

(c) Cincinnati Field Study – Distribution System

Material	Epoxy	Cement	Iron
Average log HPC (cfu/cm$_2$)	3.3	3.8	4.2

Greenville Water System

The influent water quality data for the Greenville Water System annular reactors, is shown in Table 5.5. In general, the influent pH, alkalinity, and total chlorine residual remained similar before and after the DAF treatment was implemented (see Figure 5.13). Influent HPC levels rose from early May, 2000 through mid-July, 2000, coinciding with the conversion to DAF, decreased to previous levels during the fall and winter months, and then increased again in late spring and summer 2001 (see Figure 5.14).

Table 5.5
Greenville Water Quality Data For Influent to Annular Reactors

Parameter	Before DAF	After DAF
PH	7.6 (6.8 – 8.3)	7.5 (6.3 – 9.2)
Temperature, deg C	14.6 (8.0 – 24.0)	19.0 (10.2 – 26.7)
Alkalinity, mg/L	10.8 (9.5 – 14.5)	10.4 (5.0 – 16.0)
TOC, mg/L	1.4 (1.3 – 1.5)	0.9 (0.5-1.3)
Chlorine, mg/L	2.6 (1.6 – 3.5)	3.0 (1.5-4.4)

By the end of the study, June 28, 2001, thirty-six sets of slide samples had been collected. Results of biofilm analyses of these slide samples are included in Figure 5.15, with average log HPC level shown in Figures 5.16 and 5.17 for before DAF and after DAF water quality conditions respectively. The duplicate cement and iron annular reactors did not have statistically similar attached HPC levels, there the results from these reactors were evaluated individually when making comparisons to data from the other reactors.

The iron coupons exhibited higher biofilm growth than either the cement or epoxy coupons. Table 5.6 displays the statistical comparisons of these materials. The log attached HPC levels measured on the ductile coupons before the DAF treatment was online were significantly higher than either the cement or epoxy. There was no statistical difference between the cement and epoxy coupons. All three materials had statistically different log attached HPC levels measured after DAF treatment was implemented, with iron coupons exhibiting the highest attached HPC levels. There was a statistical difference in the biofilms measured on the cement coupons before and after DAF treatment, but no difference with the iron and epoxy coupons.

Table 5.6
Statistical Comparisons of Mean Log HPC Levels for Greenville Water System. Lines between materials indicate that no statistically significant differences were found.

(a) Greenville Field Study – Before DAF

Material	Cement	Epoxy	Iron
Average log HPC (cfu/cm$_2$)	1.6	2.1	4.3

(b) Greenville Field Study – After DAF

Material	Epoxy	Cement	Iron
Average log HPC (cfu/cm$_2$)	2.2	2.5 – 3.2	4.2 – 5.3

Tampa Bay Water

The raw influent water quality data to the three reactors during the course of the study is shown in Table 5.7 below and in Figure 5.18. In general, the influent water quality was steady, with a pH of about 7.9, alkalinity of 219 mg/L, total chlorine residual of 3.4, and TOC of about 3.9 mg/L.

Table 5.7
Tampa Bay Water Quality Data for Influent to Annular Reactors

Parameter	Average	Range
PH	7.9	7.1 – 8.1
Temperature, deg C	18.4	8.0 – 26.7
Alkalinity, mg/L	219	197 – 239
TOC, mg/L	3.9	3.4-4.2
Chlorine, mg/L	3.4	2.1-3.8

A total of 14 sets of coupon samples were collected during the course of the field study in Tampa Bay. The attached HPC levels measured from the coupon are presented in Figure 5.19 and average log attached HPC values can be seen in Figure 5.20. On visual examination, there appears to be little difference between the three materials with respect to biofilm growth. Statistical comparisons are presented in Table 5.8 below, which indicates that there was no statistical difference in log attached HPC levels between the three materials tested.

Table 5.8
Statistical Comparisons of Mean Log HPC Levels for Tampa Bay Water Department. Lines between materials indicate that no statistically significant differences were found.

Material	Cement	Epoxy	Iron
Average log HPC (cfu/cm$_2$)	4.5	5.1	5.2

Comparisons of Biofilm Growth

Log attached HPC levels measured from the specific materials tested in the annular reactors were also compared between utilities. Figure 5.21 displays the log attached HPC levels for all utility studies. The iron coupons exhibited the highest attached HPC levels for all utility studies, while the lowest growth was measured on either the cement or epoxy coupons depending on the specific utility. Table 5.9 presents the statistical comparisons of the iron coupons. The highest log HPC levels were measured in Cambridge, followed by Tampa Bay. The lowest log HPC levels were measured in Cincinnati at the Miller Treatment Plant. Statistically similar biofilm levels were measured in the Cincinnati distribution system coupons, the Greenville Water System coupons before DAF treatment was implemented, and at Tampa Bay (the iron coupons from the duplicate annular reactors at Greenville after DAF treatment exhibited mixed statistical results, since the results from the duplicate iron reactors were not statistically similar and had to be evaluated separately.

Growth on cement and epoxy coupons from all utility studies was several logs lower than growth on the iron coupons. Table 5.10 presents the statistical comparison of average log HPC levels from the cement coupons. Tampa Bay exhibited the highest log attached HPC level and Greenville before DAF treatment was implemented were the lowest. Statistically similar levels were observed at Cambridge with MWRA water, Cincinnati at the Miller Treatment Plant, and

Greenville after DAF treatment was implemented, and Cincinnati at the Saylor Park Fire Station in the distribution system. Table 5.11 presents the statistical comparisons of log attached HPC levels from the epoxy coupons. Tampa Bay exhibited the highest biofilm levels on the epoxy coupons, with Greenville before DAF treatment exhibiting the lowest. Statistically similar biofilm levels were measured in Cambridge with MWRA water, Cincinnati at the Saylor Park Fire Station, and Cincinnati at the Miller Treatment Plant

DISCUSSION AND SUMMARY

Although each utility location represents a unique set of water quality conditions, the results generally support both the laboratory studies completed for this project, as well as previous studies that have demonstrated higher biofilm growth on iron surfaces, with cement and epoxy coupons exhibiting lower growth. The biofilm HPC levels measured on iron coupons in the Greenville and Cambridge field studies were higher than the biofilm HPC levels measured on cement or epoxy, however in Tampa Bay and Cincinnati, the biofilm HPC levels measured on the iron coupons were not significantly different than the cement or epoxy coupons. The cement and epoxy coupons from the Cambridge, Cincinnati, and Greenville studies all exhibited similar ranges of biofilm HPC levels, and Tampa Bay had the highest biofilm HPC levels on these materials.

While the levels of growth for any of the materials was specific to each utility field study, some general observations can be made regarding the water quality conditions for these four studies and the resulting biofilm growth observed on the coupons. Fluctuations in influent water quality occurred throughout the course of these studies, however in general, the influent water to the reactors in the Tampa Bay study had the highest TOC values (average of 4.0 mg/L C) and the highest free chlorine residual of all the studies (average of 3.4 mg/L Cl_2). The water for the Cincinnati study had the lowest TOC values (average of 0.8 mg/L C) and much lower average chlorine residuals (1.2 mg/L Cl_2). The TOC levels for the Greenville and Cambridge studies ranged from 0.9 to 2.0 mg/L C, and both of these utilities used chloramines with average values ranging from 1.6 mg/L to 3.0. No differences in attached growth were observed between materials for either the Tampa Bay study (higher TOC groundwater) or the Cincinnati study (lower TOC), whereas the iron coupons from the utilities with intermediate TOC values that chloraminate displayed significantly higher biofilm growth than the other materials (cement and epoxy) tested at each location. The biofilm HPC levels on the cement and epoxy coupons at Tampa Bay were higher than the cement and epoxy coupons from the other utility locations, indicating that the higher organic loading may have supported more growth. The growth on the iron coupons from Tampa Bay was lower than the growth on the iron coupons at Cambridge, even though the influent TOC levels at Tampa Bay were higher than in Cambridge. This might possibly indicate that the nature of the organics from the groundwater source at Tampa Bay may have been less bio-available.

Table 5.9

Statistical Comparison of Mean Log HPC levels on Ductile Iron Coupons. Lines between utilities indicate that no statistically significant differences were found.

Utility	Average Log HPC cfu/cm^2
Cincinnati – Miller WTP	2.8
Cincinnati – Distribution	4.2
Greenville – Before DAF	4.3
Greenville – After DAF	4.2 – 5.3
Tampa Bay	5.2
Cambridge – MWRA Water	7.0
Cambridge – Cambridge Water	7.0 – 7.5

Table 5.10

Statistical Comparison of Average Log HPC levels on Cement Coupons

Utility	Average Log HPC cfu/cm^2
Greenville – Before DAF	1.6
Cambridge – Cambridge Water	2.2
Cambridge – MWRA Water	2.9
Greenville – After DAF	2.5 – 3.2
Cincinnati – Miller WTP	3.3
Cincinnati – Distribution	3.8
Tampa Bay	4.5

Table 5.11

Statistical Comparison of Average Log HPC levels on Epoxy

Utility	Average Log HPC cfu/cm^2
Greenville – Before DAF	2.1
Greenville – After DAF	2.5
Cincinnati - Miller WTP	3.2
Cincinnati – Distribution	3.3
Cambridge – MWRA Water	3.3 – 4.3
Cambridge – Cambridge Water	4.2
Tampa Bay	5.1

In summary, the data collected from the four field studies generated the following results:

- The iron coupons exhibited the highest attached HPC levels, or were not statistically different from the materials with the highest levels for all utility studies.
- In general, there was no statistical difference between biofilm growth on the cement and epoxy coupons
- Iron coupons from Cambridge had significantly higher biofilm growth than the iron coupons from any of the other utility studies
- There was no statistical difference in log attached HPC level between the three materials tested at the Miller Treatment Plant in Cincinnati or in Tampa Bay.

- The attached growth measured on the cement and epoxy coupons at Tampa Bay was higher that the growth on these materials from the other utility locations. Tampa Bay exhibited the highest influent TOC values of all the utility field studies.

- The attached growth on the cement and epoxy coupons exhibited similar ranges of values in the Greenville, Cincinnati, and Cambridge studies.

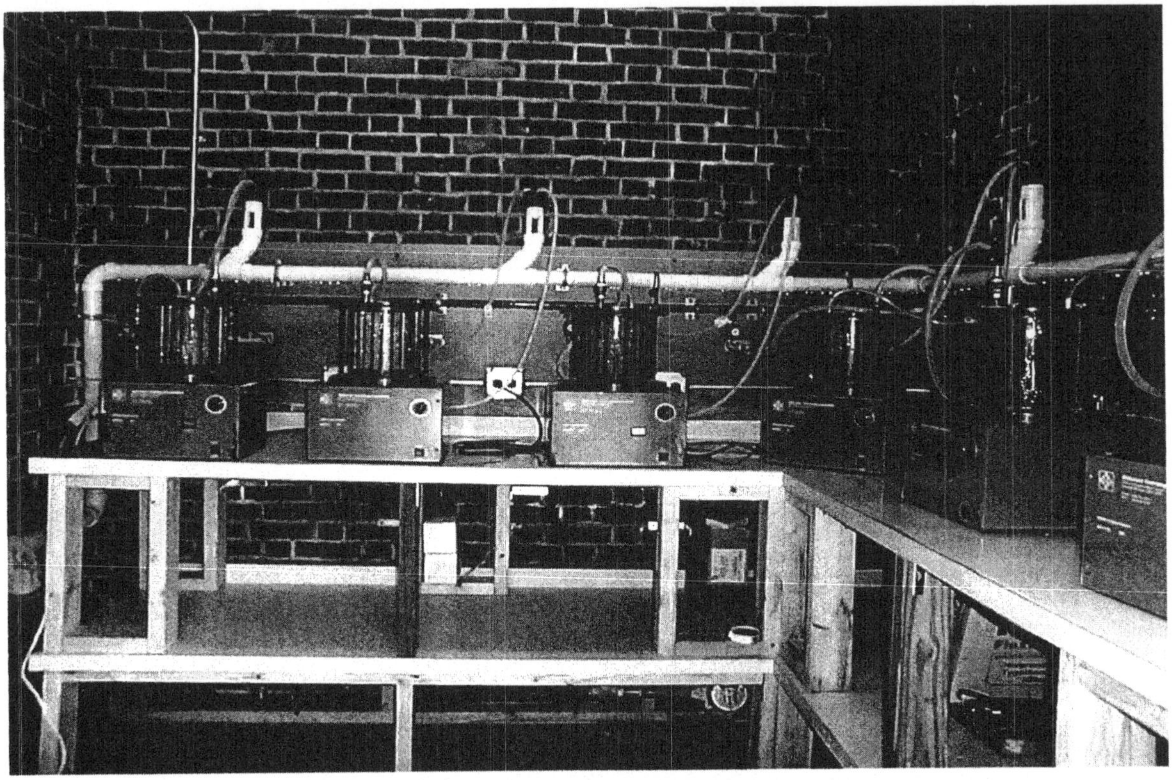

Figure 5.1 Annular reactor set-up at Cambridge Water Department

Figure 5.2 Annular reactor at Cincinnati Water Works

Figure 5.3 Inserting a new epoxy coupon into the annular reactor at the Greenville Water System

Figure 5.4 Tampa Bay annular reactor set-up at Pinellas County Water Treatment Plant.

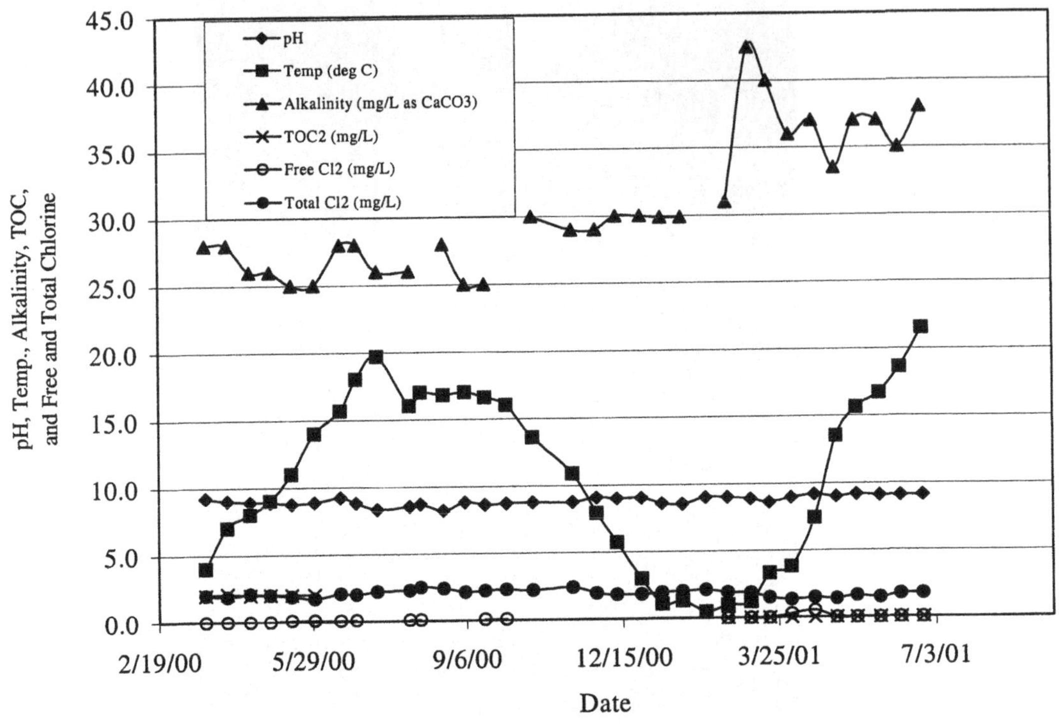

Figure 5.5 Cambridge Water Department influent water quality to annular reactors

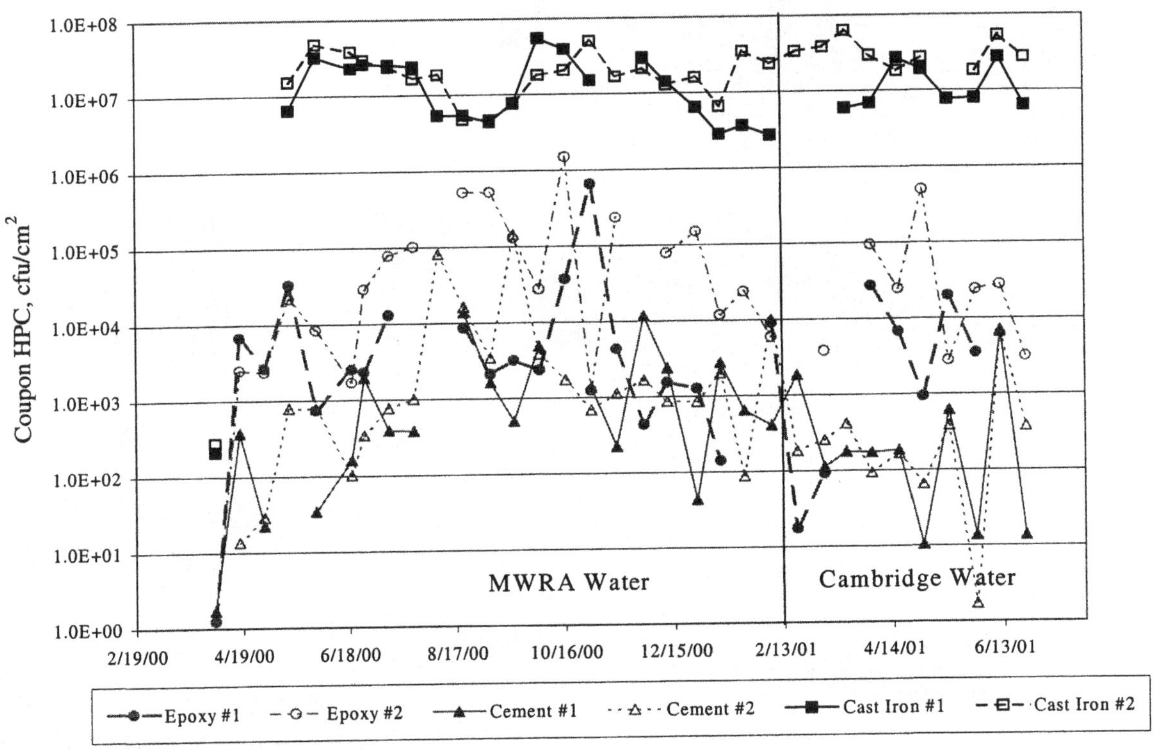

Figure 5.6 HPC biofilm densities from Cambridge, MA

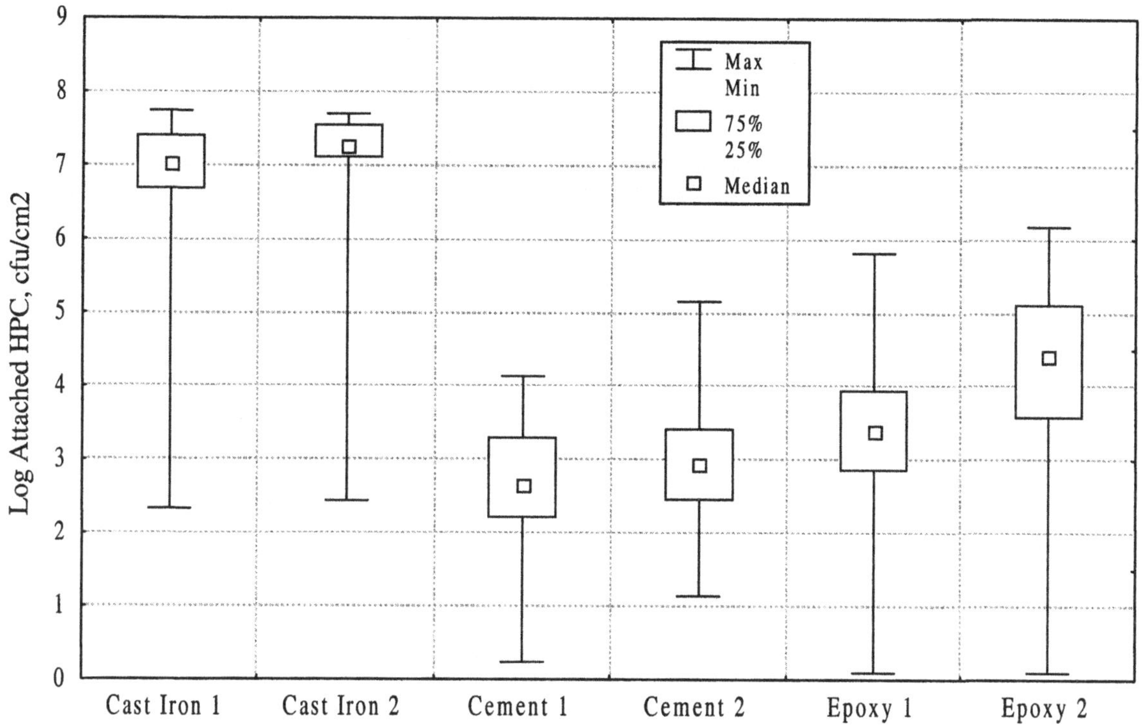

Figure 5.7 HPC biofilm densities from Cambridge, MA - March 20, 2000 through March 9, 2001

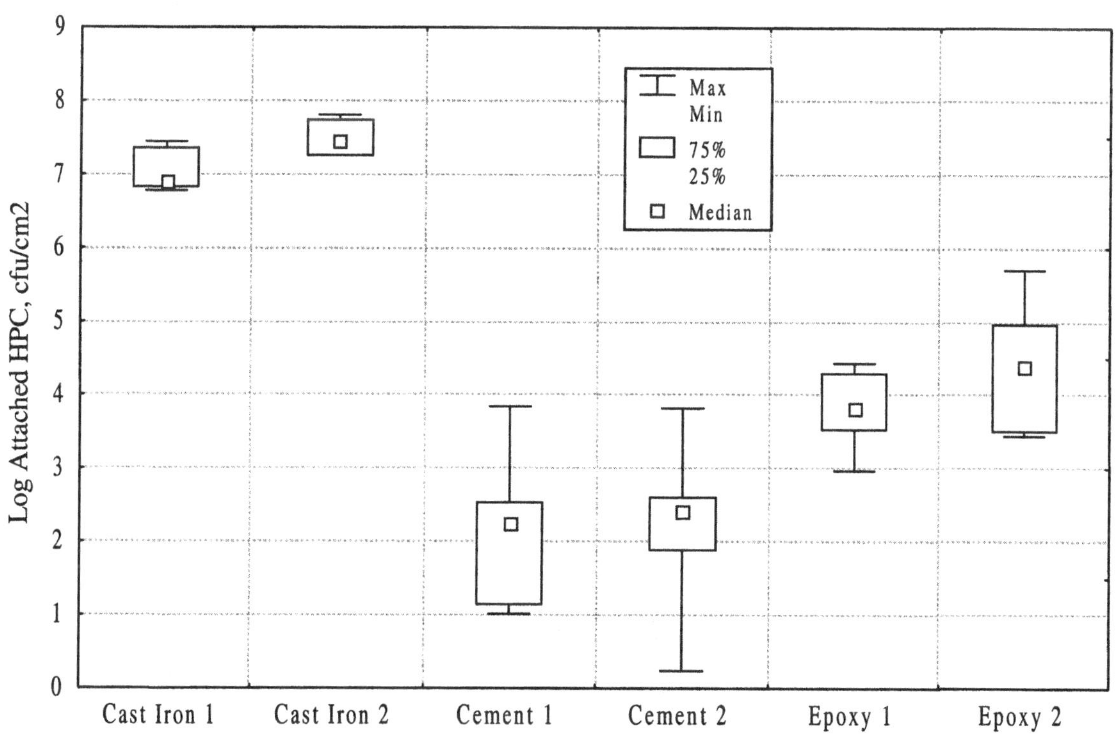

Figure 5.8 HPC biofilm densities from Cambridge, MA - March 9, 2001 through June 25, 2001

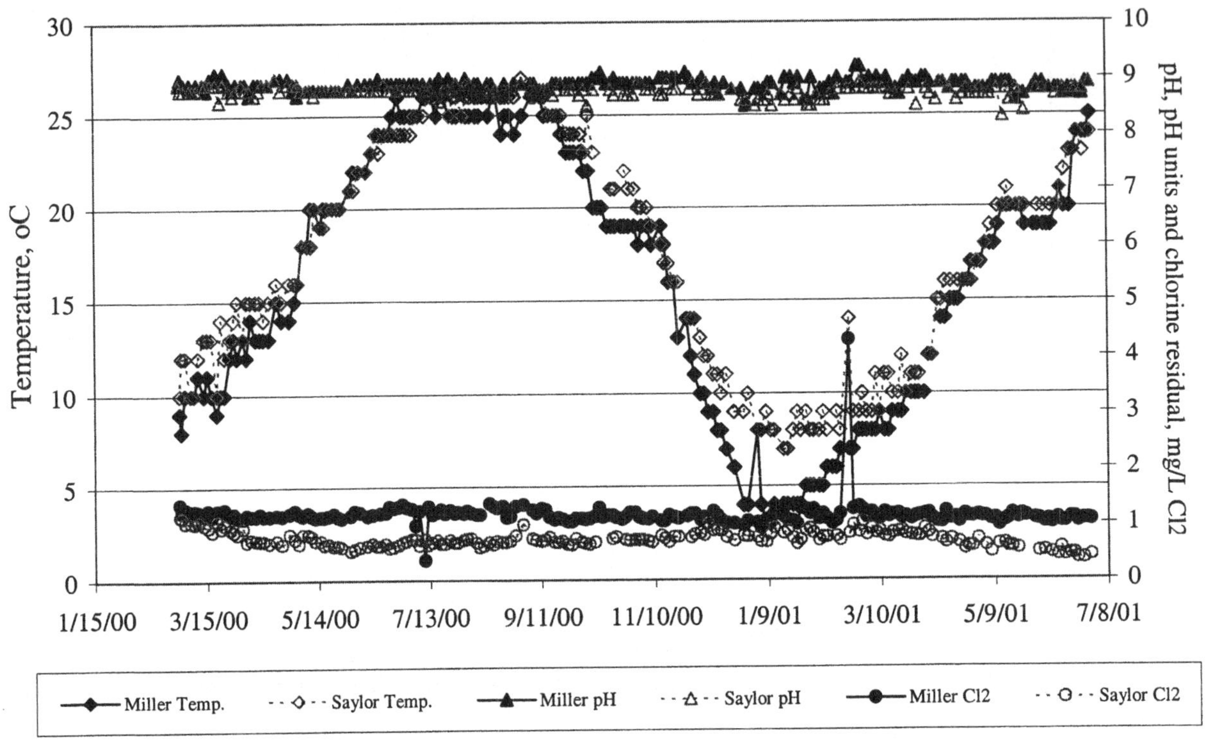

Figure 5.9 Cincinnati Water Works influent water quality to annular reactors

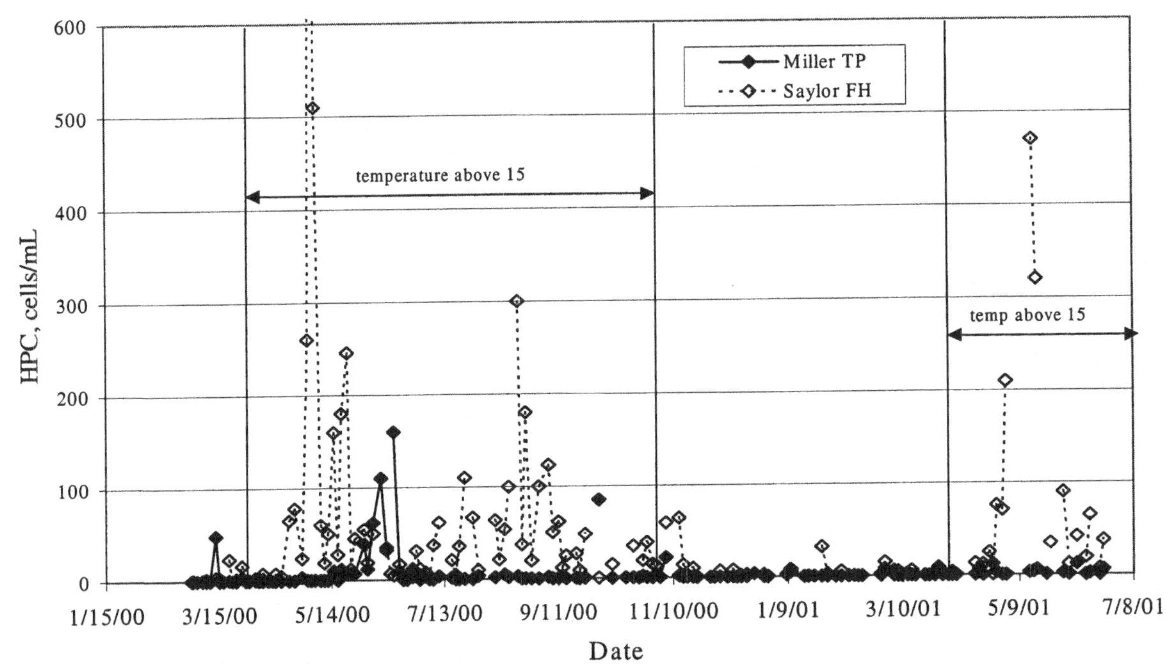

Figure 5.10 Cincinnati Water Work HPC levels in influent to annular reactors

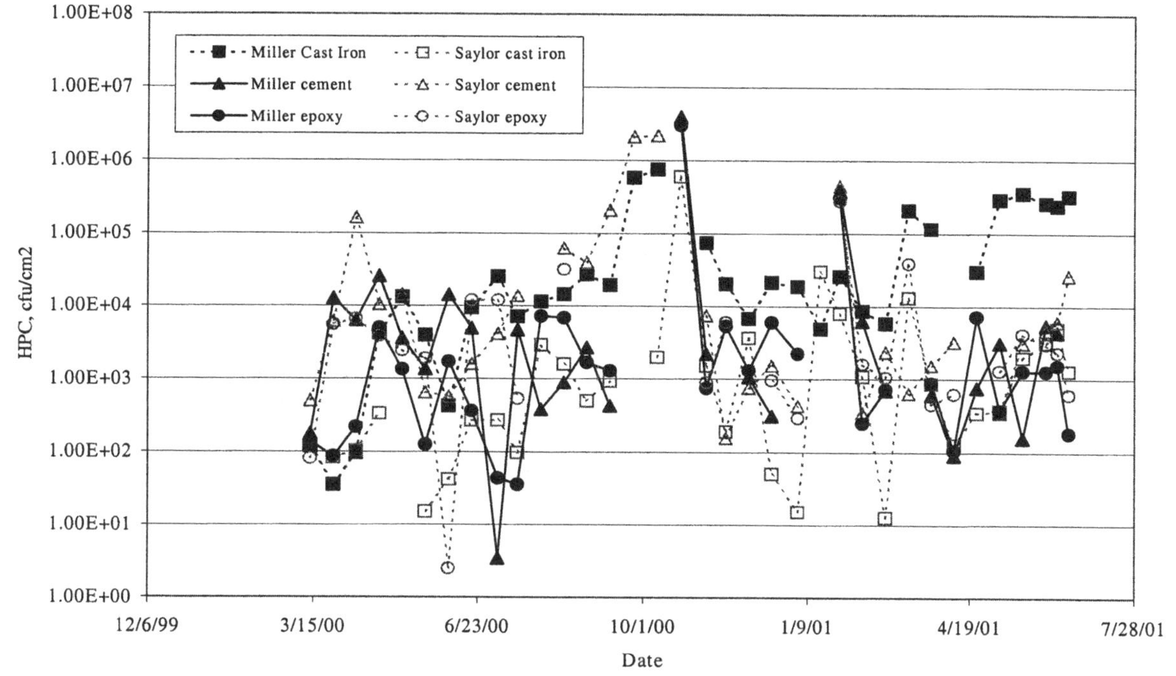

Figure 5.11 Attached HPC levels from annular reactor coupons – Cincinnati Water Works

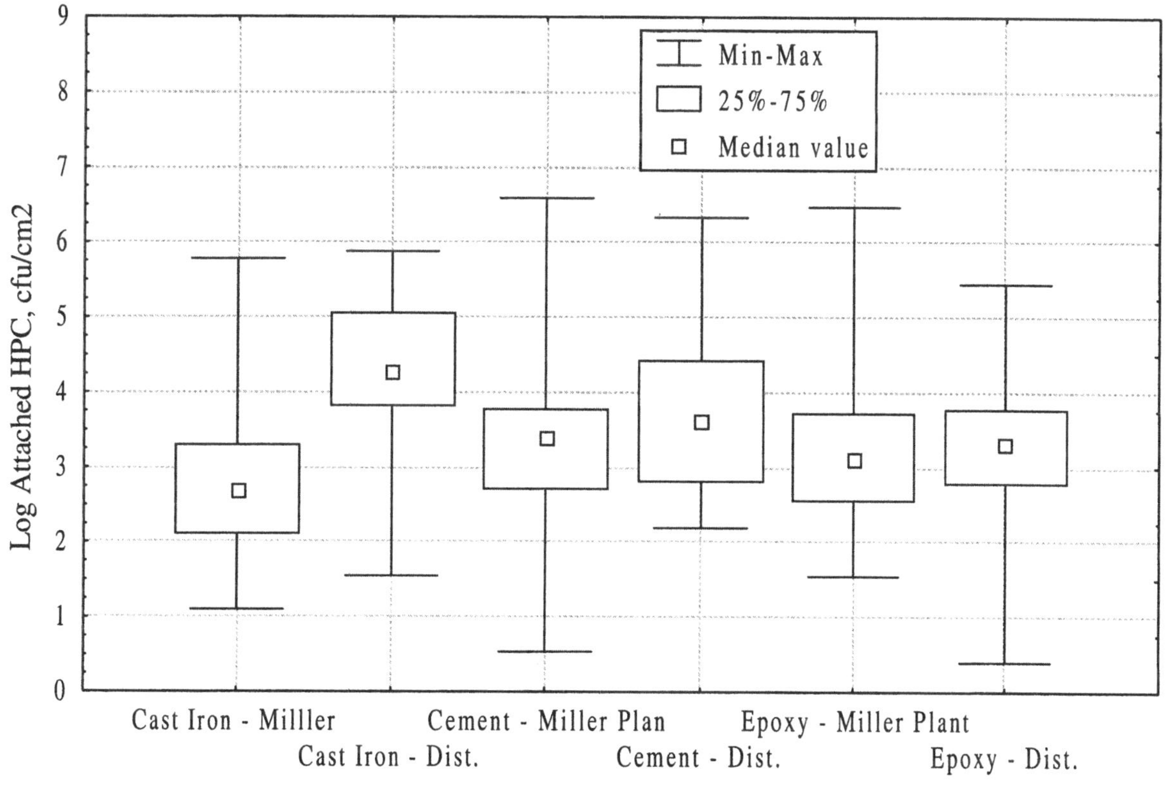

Figure 5.12 HPC biofilm densities from Cincinnati Water Works

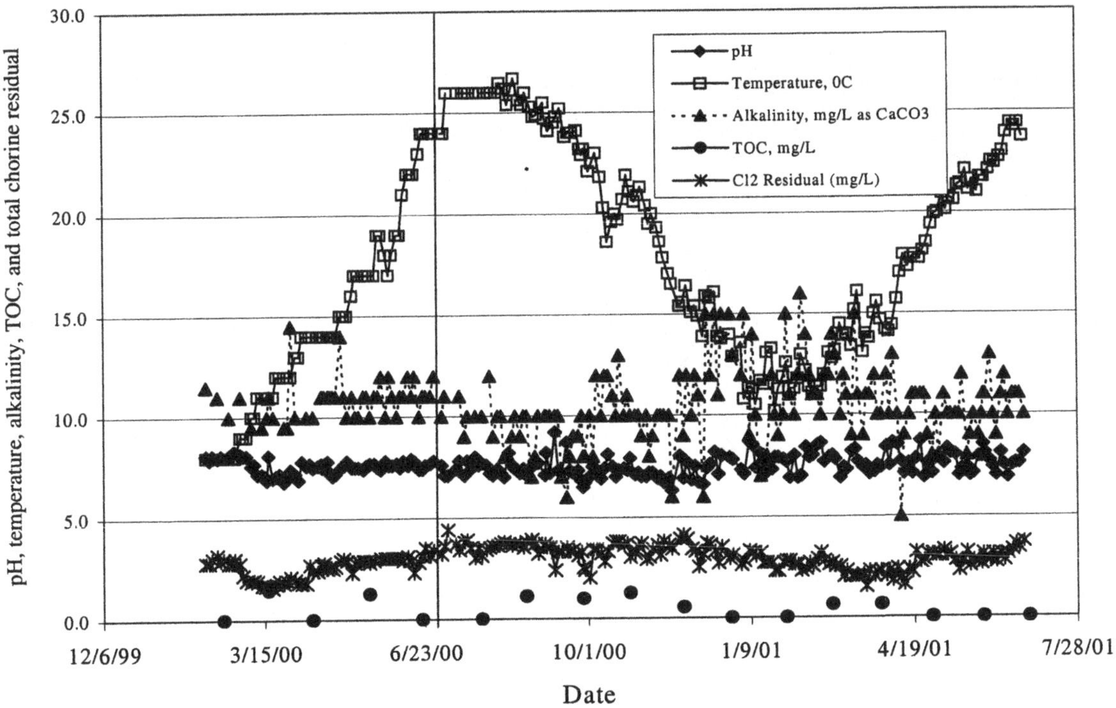

Figure 5.13 Greenville Water System influent water quality to annular reactors

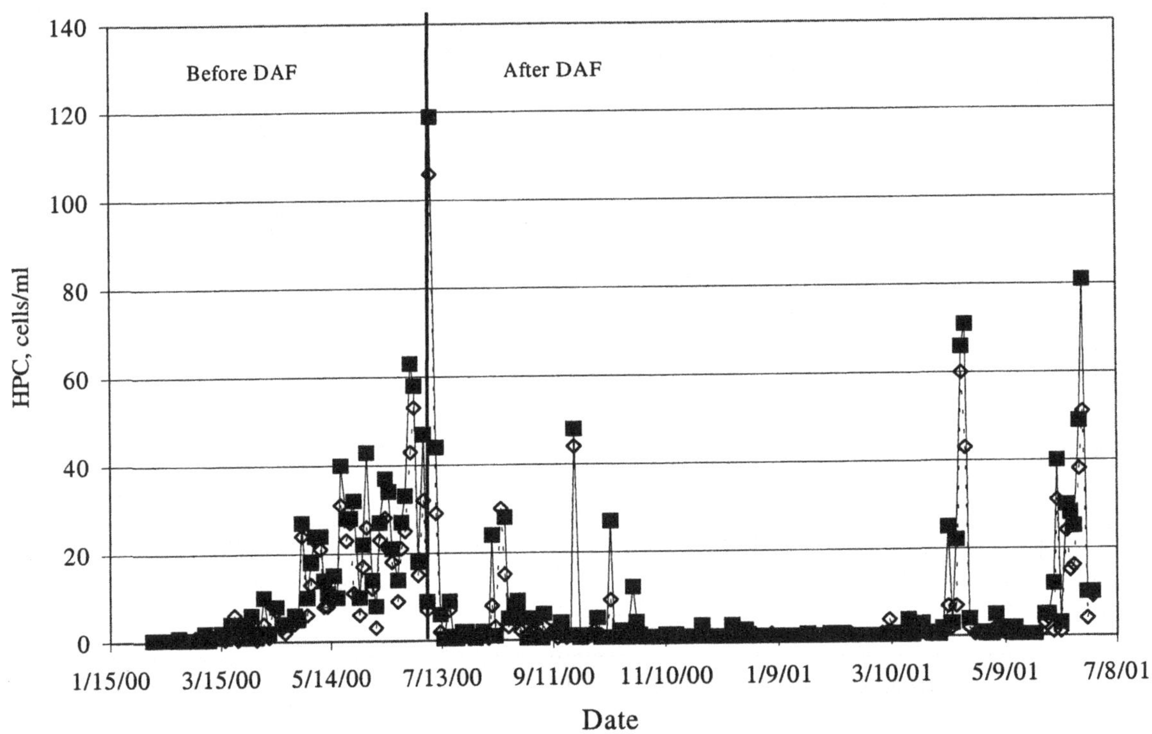

Figure 5.14 Greenville Water System influent HPC levels in influent to annular reactors

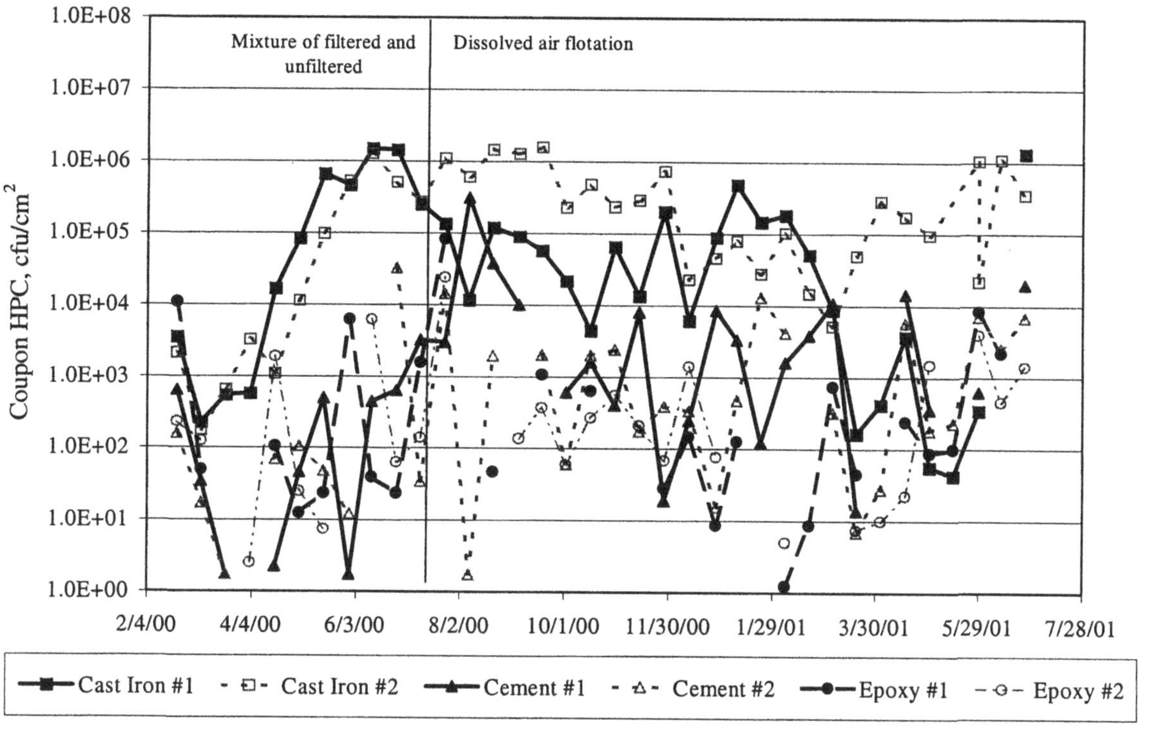

Figure 5.15 Attached HPC levels from annular reactor coupons – Greenville Water System

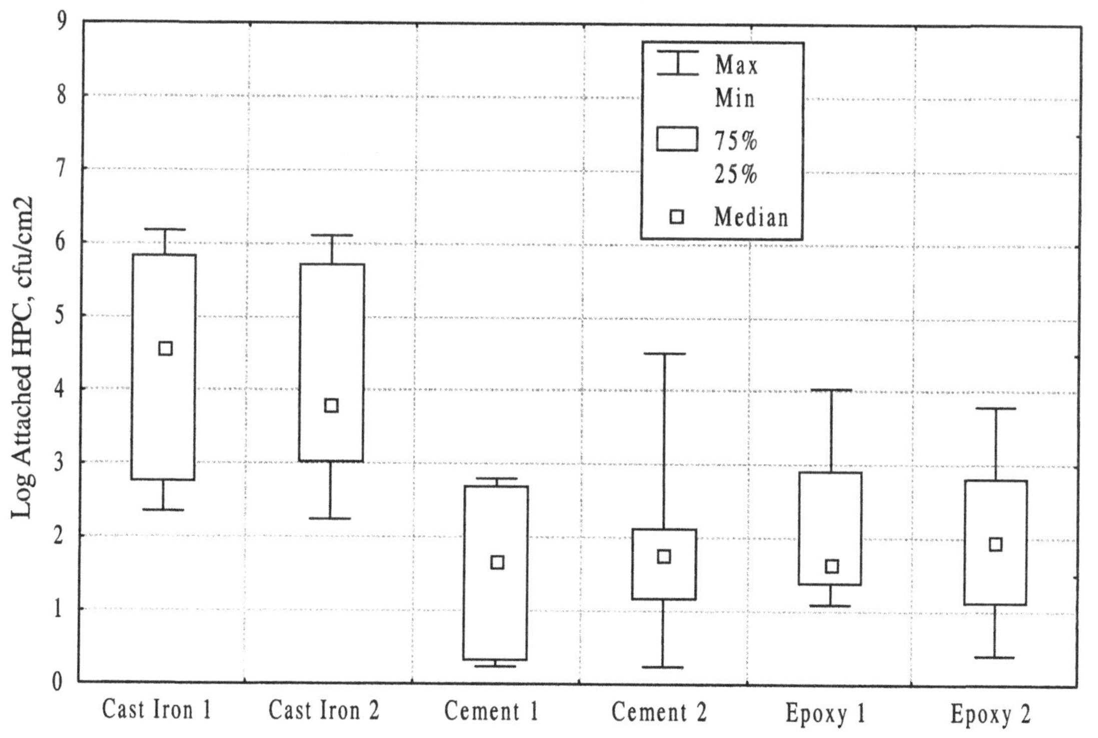

Figure 5.16 HPC biofilm densities from Greenville Water system before DAF

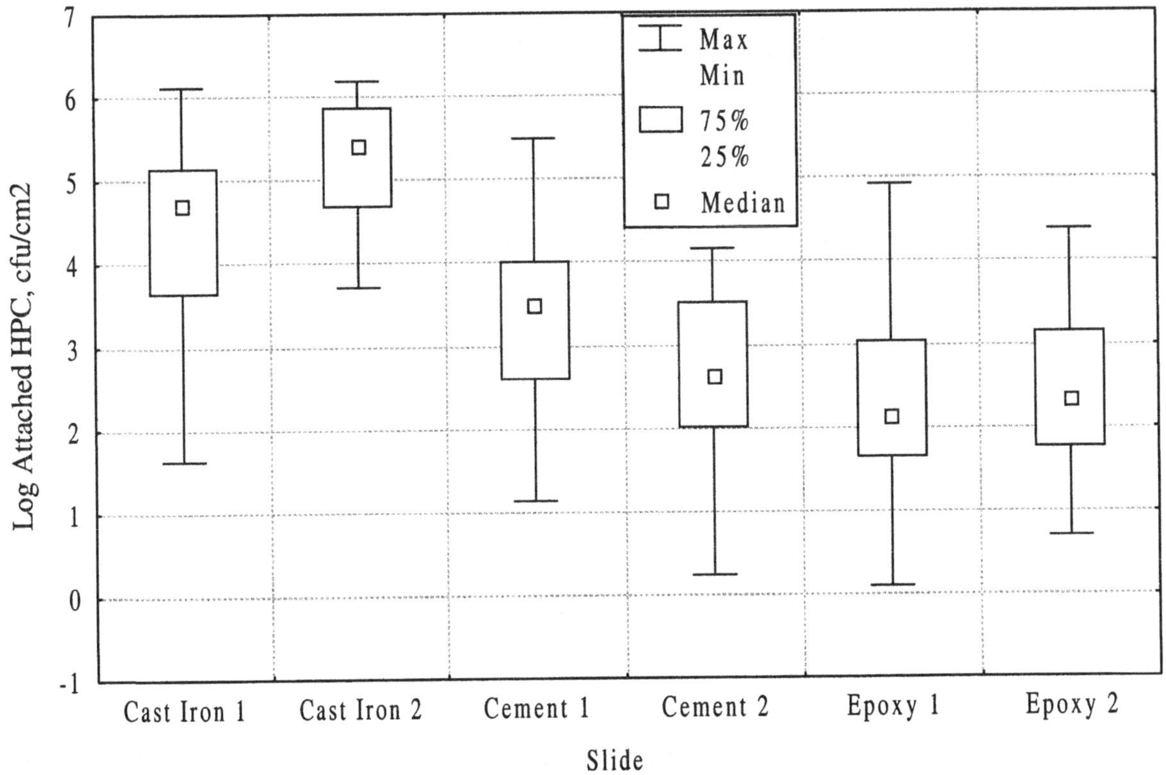

Figure 5.17 HPC biofilm densities from Greenville Water System after DAF

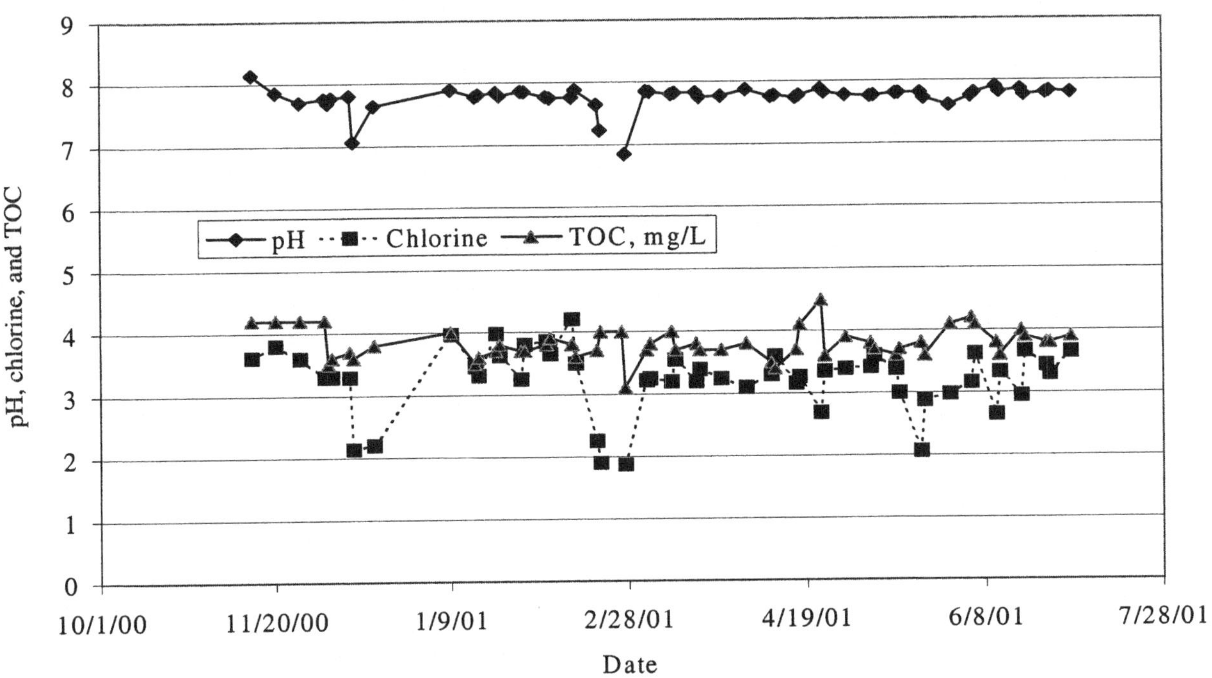

Figure 5.18 Tampa Bay influent water quality to annular reactors

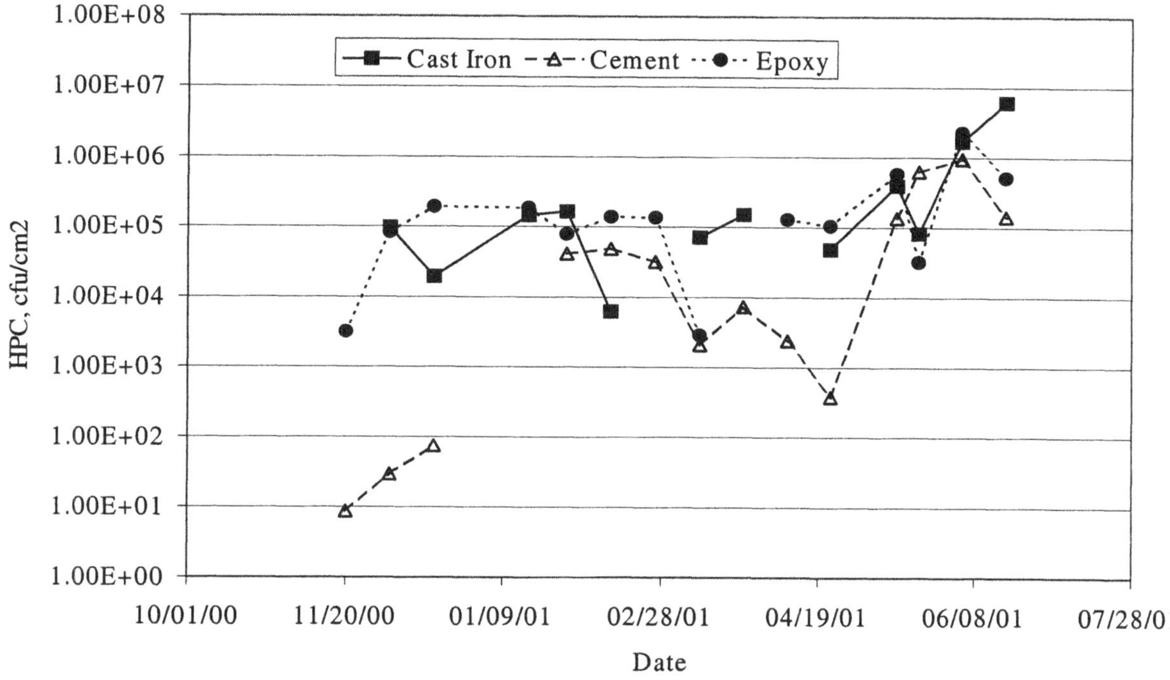

Figure 5.19 Attached HPC levels from annular reactor coupons – Tampa Bay

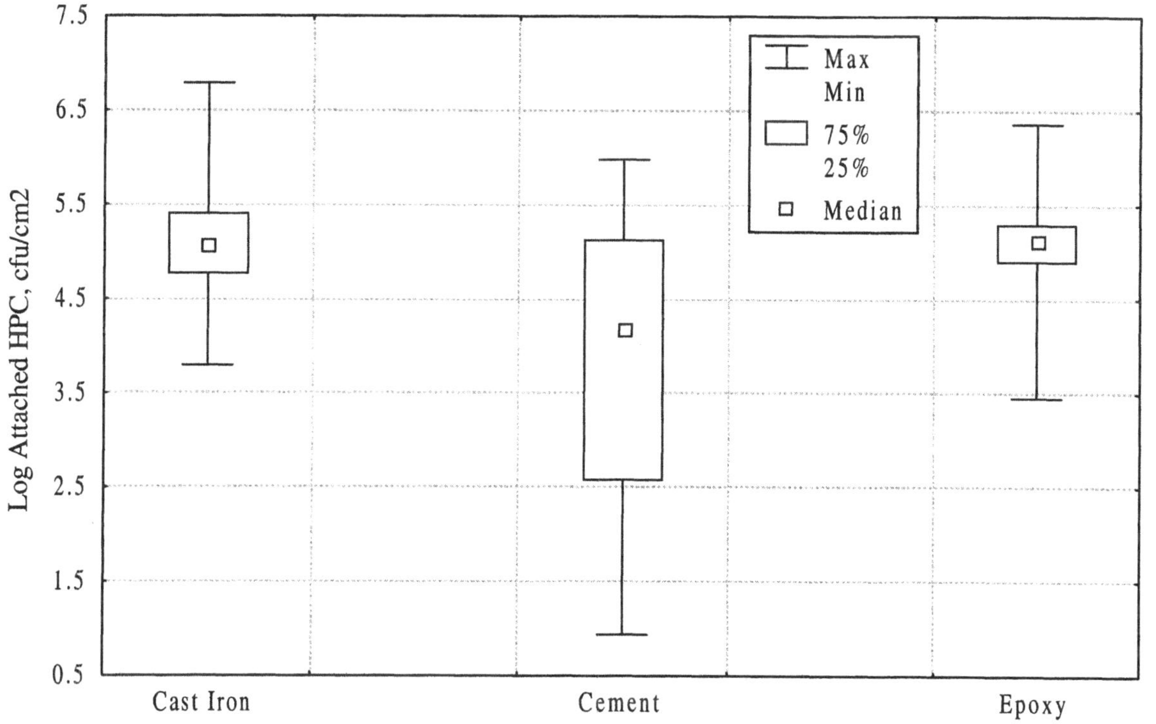

Figure 5.20 HPC biofilm densities from Tampa Bay

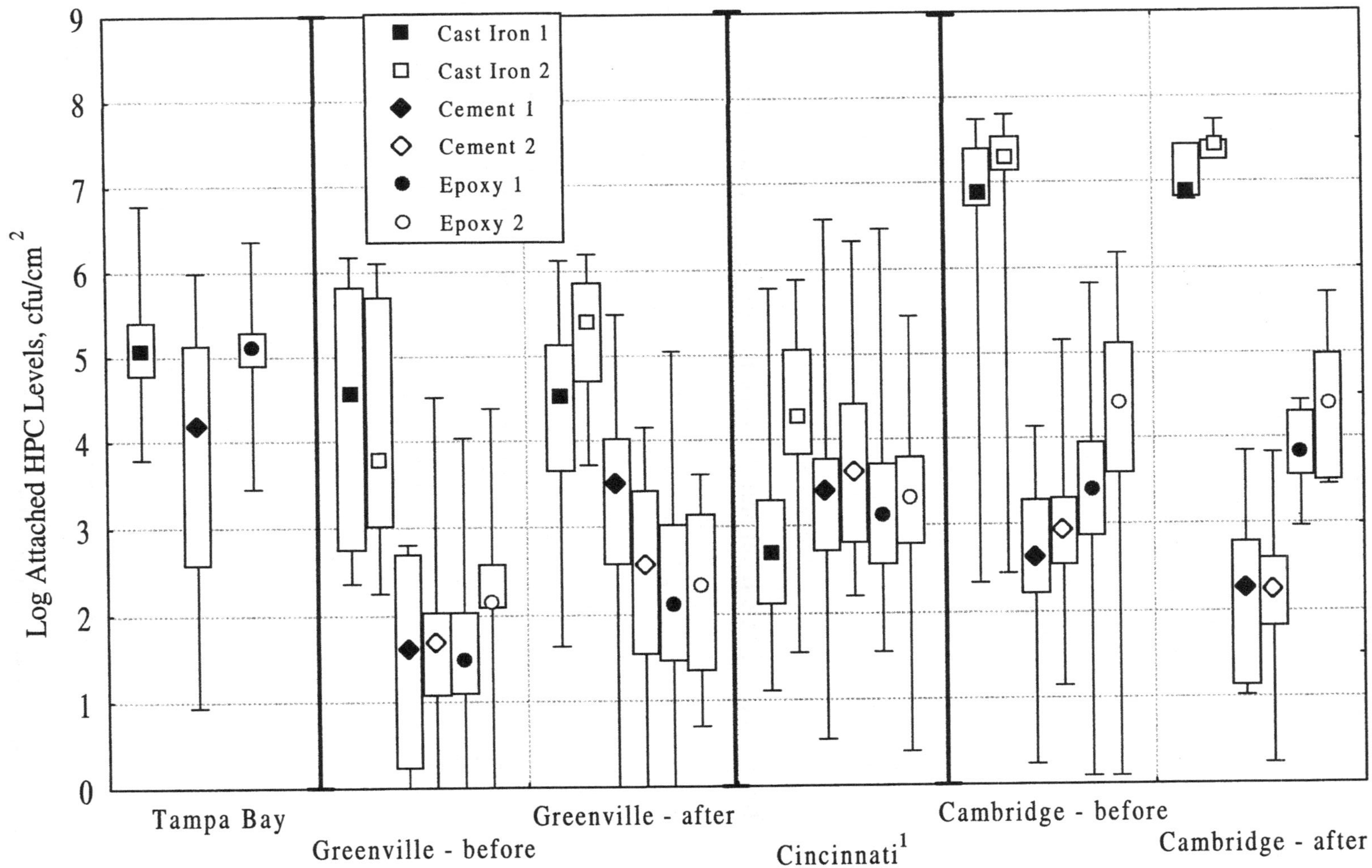

Figure 5.21 HPC biofilm densities from all utilities

[1] **(cast iron 1, cement 1, and epoxy 1 for Cincinnati are from the Miller Treatment plant; cast iron 2, cement 2, and epoxy 2 for Cincinnati are from the Saylor Park Fire Station)**

CHAPTER 6
CONTROL STRATEGIES FOR DISTRIBUTION SYSTEM REGROWTH PROBLEMS

INTRODUCTION

The aim of this research was to investigate and determine the influence of distribution system pipeline material on bacterial regrowth. This emphasis on piping material is consistent with utilities' prioritization of replacement of cast iron pipe, which can be a major contributor to water quality degradation in the distribution system. To fully understand the importance of pipe materials on regrowth, the influence of organic matter, disinfectant residual and corrosion control must be considered. While a systematic study of each of these factors was outside the scope of this work, organic matter and disinfectant type were included.

This chapter summarizes the conclusions from this study and provides guidance for utilities to assist them in understanding and controlling regrowth in their system, specifically with respect to infrastructure materials, organics, and disinfectants.

STUDY CONCLUSIONS

The laboratory testing showed that increases in carbon levels led to general increases in biofilm and planktonic densities, and this effect was most pronounced for biofilms in reactors that contained iron. In the presence of disinfectants and supplementary organic carbon, the iron reactors had the highest effluent counts of any of the materials tested. For the other materials (cement, epoxy, and PVC), there was no definite order of ascendance with respect to biofilm of planktonic growth, but the PVC reactors always exhibited the lowest levels, or weren't statistically different than the reactors with the lowest levels. The straightforward impact of these findings is that the combination of higher levels TOC and cast iron pipe can lead to relatively high numbers of bacteria.

The field testing results indicate that in certain conditions, iron resulted in the highest regrowth, and in other cases, the type of material had no influence on the number of bacteria present. Regrowth appeared to very site specific depending on the nature and quantity of the organic material, disinfectant, and type of material present. While iron pipe exhibited the highest biofilm levels, it should not be concluded that this would always be the case. Site specific testing to determine if material is a contributing factor in regrowth should be performed. This type of testing will help utilities identify if piping material; in this case iron pipe, is a primary contributor to regrowth.

The water systems in the field testing component of this study had very different levels and types of disinfectant present. The strategy that most utilities use to mitigate regrowth is to increase the disinfectant residual or convert from free chlorine to chloramine, which is considered to be more effective for controlling regrowth. Despite the very high levels of disinfectant present in this study, the system with the lowest disinfectant residual and with less effective free chlorine compared to chloramine, had the lowest regrowth on iron coupons. This system also employs GAC treatment and lowers its final distributed water TOC concentration to less than 0.80 mg/L. This is not to say that the presence of disinfectant residual does not aid in controlling regrowth, but a disinfectant residual as was present in the systems tested did not lead to lower numbers biofilm numbers on iron pipe. When regrowth is a problem, for example coliform problems in the distribution system, utilities should undergo an evaluation to determine the effects of organics and cast iron pipe.

GUIDANCE

These conclusions have practical applications to the drinking water industry. One alternative for utilities wishing to reduce biofilms in the distribution system would be to reduce the amount of organic carbon in the finished waters entering the distribution system. This study has shown that higher levels of carbon support greater biofilm growth and planktonic populations on any pipe material, although the effect is most pronounced on iron pipe. Thus, reducing the organic carbon in the finished water would be effective for any utility wishing to reduce regrowth problems in their distribution system. This can be accomplished using one of several treatment technologies, including enhanced coagulation, membrane filtration, filtration using iron-oxide coated media, or biological filtration. However, it is important to recognize that even with very low carbon concentrations, biofilms can still flourish. As observed in this study, this is especially true for unlined iron surfaces.

Utilities with significant amounts of iron pipe in their distribution systems are faced with the greatest regrowth potential. For utilities that are also distributing waters that are high in natural carbon, this problem is compounded. Thus, the best option for improving water quality may be to replace or reline the iron pipe in their distribution systems. A thorough study on the overall causes of water quality deterioration should be completed however, and because of the costs involved, utilities will have to develop a plan and budget whereby a reasonable amount of piping is replaced on an annual basis.

Although pipe replacement or relining is the best option for reducing biofilm problems, this alternative may be more economically feasible for utilities with deteriorating iron pipe that must be replaced or remedied rather than for utilities with newer iron pipes. For utilities with a smaller amount of unlined iron pipe, reducing the amount of organic carbon entering the distribution system may be a better option. However, it is important to weigh the expense of reducing the carbon in the distribution system against the actual potential for significantly reducing biofilm growth.

From a practical standpoint the findings of this study have a profound impact on how water systems make important decisions regarding regrowth. A water system experiencing regrowth problems could potentially justify a costly cast iron rehabilitation program based on results of a site specific study. On the other hand, the data from this type of study could also point to enhanced organic removal rather than removal or rehabilitation of cast iron pipe. In any case, many water systems currently are replacing cast iron pipe, with the expectation that regrowth in the distribution system should be reduced. Site specific testing may help utilities determine the most cost effective approach to solving their regrowth problems.

USING ANNULAR REACTORS FOR SITE SPECIFIC TESTING

This study placed annular reactors at four different utilities for a period of 6 to 18 months. There were operational problems with the annular reactors, but the data produced were consistent and reliable over time. If a water system wishes to investigate the cause of regrowth within their distribution system, annular reactors when operated correctly, with analyses completed by a qualified lab, can be a practical tool to accomplish this. Therefore, it is very important for utilities to realize that annular reactors are more than just a research tool, but can be used as a reliable tool to assist utilities in making decisions.

Careful consideration must be given to the study design, including the location of the reactors, flowrate, and treatment conditions. When conducting an annular reactor study to determine if a treatment change (i.e. organic removal) is likely to have an impact on regrowth, sufficient time must be allowed both before and after the treatment change. This is due to the slow rate of chemical and biological changes at the pipe surface. Therefore, conducting annular reactor studies for 6 - 12 months before the treatment change, followed by another 6 –12 months after the treatment change may be necessary.

SUMMARY

In summary, the main points of guidance to water systems are:

- In certain cases the combination of natural organic matter and iron pipes could be the cause of a bacterial regrowth problem. Steps should be taken whenever there is a defined regowth problem, to determine the cause.
- Tools are now available to make an assessment of whether iron pipe is the likely cause of the regrowth problem.
- If bacterial numbers are very high for iron pipe, two options may be viable. Depending on the outcome for reducing regrowth, these options are 1) removal or relining of unlined iron pipe, or 2) reduction of organic matter.

APPENDIX A
PRELIMINARY AND IN-DEPTH UTILITY SURVEYS

INITIAL WATER DISTRIBUTION SYSTEM SURVEY

Utility Name____
Your Name____
Address____
City, State, Zip____
Phone Number____
Fax Number____
E-mail____
Date____

General Characteristics of the Water System

1. Where do you get your source water? Please list all sources.
2. What is the population served by the distribution system?
    a. <10,000
    b. 10,000-50,000
    c. 50,000-100,000
    d. 100,000-500,000
    e. >500,000
3. How many municipalities does the distribution system serve?
4. What is the average daily demand of the water system?

Treatment Process

1. How would you characterize your treatment scheme?
    a. Conventional (coagulation, flocculation, sedimentation, filtration)
    b. Direct filtration
    c. Other (specify)
2. What method do you use for primary disinfection?
    a. Chlorine
    b. Chlorine Dioxide
    c. Potassium Permanganate
    d. Ozone
    e. Ultraviolet Light
    f. Other (specify)
3. What method do you use for secondary (distribution system) disinfection?
    a. Free Chlorine
    b. Chloramines
    c Other (specify)
4. What is the target disinfectant residual leaving the plant?
   What is its typical range in the distribution system?
5. What type of corrosion control do you use?
    a. Orthophosphate
    b. pH adjustment (to what pH level?)
    c. Other (specify)

Coliform Levels

1. How many times have you exceeded the 5% positive coliform level in the past 10 years? What were the dates of the exceedences?
2. Was the presence of coliform bacteria attributed to a single factor or a combination of factors? Please list the factors
3. What evidence was collected to support this finding?
4. Have you initiated or plan to initiate any treatment, operations, or modifications to overcome your coliform problems? Please list the methods you use/plan to use.

Water Quality Data

1. How often do you test for the following in the distribution system?

Disinfectant residuals	Daily	Weekly	Monthly	Never
pH	Daily	Weekly	Monthly	Never
iron levels	Daily	Weekly	Monthly	Never
HPCs (Standard or R2A?)	Daily	Weekly	Monthly	Never
Coliforms	Daily	Weekly	Monthly	Never

Condition of the Distribution System

1. How old are the majority of pipes in the distribution system?
   How old is the oldest pipe?
2. What is the total volume of storage in the distribution system?
3. What percentages of the pipes are unlined cast iron?
4. What percentage of the pipes are other materials?

Material	Percentage
___________	________
___________	________
___________	________

5. What maintenance and cleaning procedures are performed? How often are these done?

DETAILED SURVEY QUESTIONS

Water Quality

1. What percentage of your water comes from the following sources?
   % surface water________ % groundwater_________ % purchased________

2. What type of corrosion control do you use?______________________________
   At what concentration?____________

3. Have fecal coliforms or E. coli bacteria ever been detected in the distribution system? If so, please complete the following table.

Percent of Total Coliform	Number of Times in the Past 10 Years	Dates of exceedences (Month and Year)
Fecal Coliform____%		______________
		______________
		______________
E. Coli______%		______________
		______________

4. Did you have to make treatment changes to respond to any coliform incidents?______

5. Has your utility ever applied for or been granted a Total Coliform Rule Variance? Is one in effect now?_______________________________________

6. Does your utility have a program for rehabilitation or replacement of its unlined cast iron pipe? If yes, please describe._____________________________

7. Do you have a formal procedure for tracking and addressing customer water quality complaints?_________________________________What is the most common type of complaint?_______________________________
8. How often do you test for turbidity in the distribution system? ______________

Sampling locations
1. Are there certain sampling locations that seem more prone to incidence of coliform positive results?____
2. What types of pipe material and service line material are most common at these sampling locations?________________________________
3. Are many of these sites close to each other or spread out throughout the distribution system?
   ____

4. What is the typical water age for sample sites in the system?___________________ What is the maximum water age?____

Distribution System Maintenance

1. Do you regularly flush your distribution system? _____ If yes, how often?______ Do you use unidirectional flushing?____ What pipeline velocities do you target during flushing? ____________________
2. Do you find that it takes longer to clean up water within some pipe material types, such as unlined cast iron, than other materials when you flush? ________________
3. If your system does not flush, why not?_____________________________________
4. Do you have a long term maintenance and/or capital improvement plan that guides distribution system pipe replacement, tank maintenance, and other infrastructure work? Please describe.____
   ____

   Specifically, does this plan include a program for rehabilitation or replacement of its unlined cast iron pipe? If yes, please describe. ______________________________
   ____

5. What procedures do you undertake to place a main in service after a replacement or break? Re-chlorination? Local flushing? Collection of coliform samples?_____________________
   ____
6. Do you have a hydraulic model of your system? If so, what do you use the model for?
   ____
   ____
7. When pipe failures or breaks occur, what do you observe as contributing factors?
   Pipe age?___
   Corrosion of pipe material?___
   Inadequate depth of bury of pipe?___
   Pipe subject to freeze-thaw?___
   Pipe subject to heavy loads?___
   Pipe subject to pressure transients?___
   Other factors?____

8. What materials do you have for water storage tanks?

Coated steel or iron___
Concrete___
Open, earthen reservoirs___
Other__________________

APPENDIX B
CASE STUDIES

CITY OF RICHLAND, WASHINGTON

Introduction

The City of Richland, WA gets 90% of its water supply from a surface water (Columbia River) and 10% of its supply from wells. The population served is 36,860, which is all one municipality. The average daily demand of the water system is 20.27 MG and the total volume of storage in the system in 24 MG.

The treatment process is direct filtration with chlorine for primary and secondary disinfection. The target residual level leaving the plant is 1.0 mg/L, while the range in the distribution system is 0.3-1.0 mg/L. No corrosion control is used.

The City of Richland was originally founded as a government town along with the Hanford Nuclear site during World War II. Since the founders didn't expect anyone to live there beyond 10 years, only thin-walled steel pipes were used in the distribution system. As Richland grew into a civilian town, other types of pipes were added. The steel pipes are slowly being replaced. The Richland distribution system is comprised of approximately 300 miles of pipe mains, ranging in size from 36 inches to 8 inches. The pipe material breakdown is as follows:

- Asbestos Cement – 38%
- Steel – 35%
- Cast Iron – 8%
- Concrete – 6%
- Unlined Cast iron – 6%
- Unknown – 4%
- PVC – 3%

The majority of the asbestos cement pipes are 20 years old. The oldest piping in the distribution system is the 50-year old steel piping.

Richland was chosen for a case study because it has a direct filtered surface water source, a small percentage of unlined cast iron pipe in a relatively new distribution system, and no coliform incidences.

Water Quality

Richland operations personnel sample chlorine residuals daily, while heterotrophic plate count bacteria and coliforms are tested for weekly. Iron and pH are not sampled in the distribution system. Chlorine is added again at two of the pump stations in the distribution system. Leaving the plant, typical water parameters are as follows:

Table B.1
Typical water quality leaving the plant – Richland, Washington

Parameter	Minimum	Maximum	Average
Turbidity	0.03 NTU	0.06 NTU	0.045 NTU
PH	6.7	7.9	7.3
Alkalinity	33 mg/L	60 mg/L	47 mg/L

The sampling ports are spread throughout the system in areas of steel pipe, asbestos cement and other distribution mains, along with galvanized steel and copper service lines. The typical water age for the sample sites was reported as 24 hours and the maximum age was 48 hours. If any parameter is outside the normal range, re-testing is done. About once a year, a positive HPC or coliform sample occurs, but the follow-up tests are always negative.

Operations and Maintenance

Richland does not have a systematic flushing program. On average, the dead ends are flushed once a year, which usually occurs when a main is replaced or repaired after a break, during installation of new mains, on dead end lines, and as a response to water quality complaints. Flushing water from the unlined cast iron pipe takes longer to clean up than with other pipe materials.

Richland samples new pipe and pipe break repairs very carefully before putting a main in service. A set of coliform samples is taken prior to pressure testing and initial flushing. After pressure testing and flushing, another set of coliform samples is collected and analyzed and only when no coliform bacteria are found is the pipe placed in service.

Since such a small percentage of the pipes are unlined cast iron, there is no plan for rehabilitation or replacement of this pipe. Richland's long term maintenance plan consists of replacement of pipes based on leaks, which are usually attributed to pipe age and material. Since there is such a large number of old, thin-walled steel pipes, Richland has been known to lose up to 30% of their water to leaks. In 1998, there were 200 main leaks, but due to regular maintenance and replacement of mains, they were down to 133 leaks in 1999.

The maintenance program also consists of cleaning of tanks. Half of the storage tanks are cleaned and inspected each year so that every tank is maintained on a two year cycle. The storage tanks are made out of coated steel or concrete.

Richland also has a formal customer complaint tracking procedure. When a complaint is called in, it is categorized and baseline information about the location, timing (intermittent or chronic), and particular customer observations are noted. When a water quality complaint is registered, Richland sends people to take chlorine, HPC, and coliform samples at the site. The results are then filed and mailed to the customer. The most common type of complaint is rusty or discolored water and the second most common is taste and odor. There is a seasonal pattern to the water quality complaints; the majority come in the fall. As of October 2000, Richland had received around 70 complaints, with 40 of them coming in the month of September alone, due to algal blooms in the Columbia River.

CITY OF REVERE, MASSACHUSETTS

Introduction

Revere, MA is one of the 60 suburbs of Boston that purchases its water from Massachusetts Water Resources Authority (MWRA). The population served by the Revere distribution system is 47,700. The average daily demand of the water system is 4.3 MGD.

The City of Revere receives unfiltered surface water that the Massachusetts Water Resources Authority has treated with free chlorine as a primary disinfectant and chloramines as the secondary disinfectant. The water is fluoridated and receives pH and alkalinity adjustment using sodium carbonate and carbon dioxide to render it less corrosive before MWRA provides it to wholesale customers such as Revere. The City does no further treatment of the water that it purchases.

Revere has a 100 mile pipe network, with pipes ranging in size from 16 inch mains to ¾ inch service lines. The majority of the pipes in the distribution system are 50-80 years old, with 70% of them unlined cast iron. Revere was chosen for a case study because it is served from a large unfiltered surface water source, has a high percentage of unlined cast iron pipe in the distribution system and has had one acute case and one chronic case of coliform exceedences in the past nine years.

Water Quality

The MWRA takes monthly samples at the entrances to the City's distribution system to test for disinfectant residuals, pH, iron levels, HPCs, coliforms, and turbidity. The City of Revere tests at thirteen sampling sites weekly for chlorine residuals, pH, iron levels, HPCs, and coliforms. The sampling sites are spread throughout the city and are located where most of the service lines are made of copper.

The normal level of residual chlorine is 1.5 mg/L, normal pH is 8.7, normal iron is 26.7 μg/L, and normal HPC is 188 cells/mL. If any parameter is outside the normal range, MWRA notifies the City of Revere of the results of the test and works with the City to come up with a solution. The City of Revere sends a yearly water quality report out to its customers.

Revere had one period of coliform exceedance from June through October, 1995, when several of the neighboring towns were experiencing similar problems. Fecal coliforms and *E. coli* were detected and a boil order was issued. The coliforms seemed to be isolated in one section of the City, where there was a mixture of unlined cast iron, ductile iron, and new pipe. These sections did not have lower than normal chlorine levels (around 1.5 mg/L). There were dead ends that probably contributed to the isolation of the coliforms. The coliforms were sometimes isolated in just one section of other MWRA towns as well. Revere dealt with the coliform problem by setting up temporary chlorination stations and flushing the lines where the coliforms were isolated. Revere also had one coliform violation in 1997. The incident was confined to one location so Revere did not have to resort to temporary chlorination stations for this incident. MWRA assisted in the treatment of sites until coliform levels were within regulatory limits.

Operations and Maintenance

The distribution system is flushed once every other year, using unidirectional flushing at the maximum velocity possible. In between the system wide flushing events, local flushes are performed sporadically at "hot spots" or on an as-needed basis. Flushing water from the unlined cast iron pipe takes longer to clean up than with other pipe materials. Flushing also occurs when a main is replaced or repaired after a break. Coliform samples are also taken at this time. Contributing factors to breakage include pipe age and ground movements.

Revere has one 2 million gallon storage tank. The tank is inspected every two years and was recently relined. Revere has a long-term plan for relining and replacing its old pipes using loan money from a program that MWRA has established with their member communities. The amount replaced per year depends on the money available at the time. In 1999, 2-3 miles of pipe were cleaned and relined and ¼" mile of new pipe was installed. Revere also opened some of the dead ends and made loops. History of the pipe and coupon samples determine the priority a pipe has in getting replaced or relined.

Revere also has a formal customer complaint tracking procedure. Complaints are called into one of two offices in the city. The two lists are not compiled into any sort of database, but each complaint is addressed individually. Revere sends someone out to the complaint site to test for potential problems. The most common complaint is odor, which is due to high chlorine residuals. Other complaints include discoloration, which is a result of on-going construction disturbing the soil and pipes.

GREENVILLE, SOUTH CAROLINA

Introduction

Greenville Water System obtains 100% of its water from surface water sources: North Saluda Reservoir, Table Rock Reservoir, and Lake Keewee. The population served by Greenville is 320,000, which includes 12 municipalities. The average daily demand of the water system is 55 MGD.

Prior to July 4, 2000, the water in the system was a mixture of conventionally treated water from the lake and unfiltered water from the two reservoirs. Now, all of the water from both reservoirs is fed to a new dissolved air flotation (DAF) filtration plant. All plants use chlorine for primary disinfection and chloramines for secondary disinfection. The target disinfectant residual level leaving the plant is between 3.0 and 3.5 mg/L and its range in the distribution system can be anywhere from 1.0 to 3.5 mg/L. Zinc polyphosphate at a concentration of 1.2 mg/L is used for corrosion control.

Water storage tanks are made out of coated iron or steel and concrete. Total storage capacity of the system is 77 MG. The length of the distribution system is 2150 miles and the largest and smallest diameter pipes are 72 inches and 1 ½ inches respectively. The pipes in the distribution system are composed of the following materials:

- Unlined Cast iron – 20-25%
- Galvanized – 15%
- PCCP Concrete – 4%
- Lined cast iron/ductile – 55-60%

The oldest pipe in the system is 110 years old. Forty-two percent of the pipes in the system are older than 27 years, 26% of the pipes are 20 to 26 years old, 16% of the pipes are 10 to 20 years old and 16% of the pipes are less than 10 years old.

Greenville was chosen for a case study because of its combination of unfiltered and filtered sources, its small amount of unlined cast iron pipe in the distribution system, and a high number of coliform incidences in the past nine years. For the purposes of this study, high coliform incidence is defined as multiple incidences over more than one year of the presence of coliforms at each of the three levels of concentration outlined in the survey form, 0-2.5%, 2.5-5%, and >5%.

Water Quality

Samples are taken daily to test for disinfectant residuals, pH, turbidity, HPCs, and coliforms. Iron levels are tested monthly. The sampling sites are spread throughout the city. Greenville reported no locational pattern to coliform positive samples. The typical water age at these sites is 3 days and the maximum water age is 5 to 6 days. Table B.2 presents typical water quality leaving the plant.

Table B.2

Typical Water Quality leaving the plant – Greenville, South Carolina

Parameter	Minimum	Maximum	Average
Turbidity, NTU filtered	0.05	0.35	0.10
Turbidity, NTU unfiltered	0.35	2.5	0.90
pH	6.8	10.2	7.5
Cl_2 Residual, mg/L	0.05	3.5	2.8
HPCs	<1	>5700	80

Greenville has had several positive coliform incidents in the past 9 years, including 6 cases of Total Coliform Rule exceedences between 1992 and 1994. They also had 10 cases of fecal coliform presence and one case of E. coli positive samples confirmed between 1991 and 1995. Greenville reported 92 incidences of coliforms in the 0-2.5% range, 10 incidences of coliforms in the 2.5-5% range, and 6 incidences in the >5% range over the past nine years. Table B.3 presents an outline of the timing and severity of coliform presence.

There is a distinct seasonal pattern to the higher levels of coliform incidence. The warmer water months of summer and early fall tended to have the highest coliform incidence.

The presence of coliform bacteria was attributed to low total chlorine residuals (less than 2.0 mg/L) and lack of an organized flushing program. Daily chlorine test data helped support this conclusion although no correlation between chlorine residual and coliforms was found. Sites with low chlorine residual were not more prone to coliform isolation. The chlorine residual above which coliforms were not found was generally not above a total chlorine residual of 2.0 mg/L, but coliforms have been isolated with chlorine residuals as high as 3 mg/L (25% of isolates between 1992 and 194 had a residual above 3 mg/L total chlorine).

Table B.3
Coliform incidents in Greenville

Year	Percentage exceedences		
	0-2.5%	2.5-5%	>5%
1991	Jan, Feb, Mar, Apr, May, Aug, Sept, Oct, Nov, Dec	June July	
1992	Jan, Feb, Mar, Apr, Sept, Oct, Nov, Dec	August	May, June, July
1993	Jan, Feb, Mar, Apr, May, June, July, Aug, Nov, Dec	October	September
1994	Jan, Feb, Apr, Oct, Nov, Dec	Mar, May, Aug, Sept	June, July
1995	Jan, Feb, Mar, Apr, May, July, Aug, Sept, Oct, Nov, Dec	June	
1996	Jan, Feb, Mar, Apr, May, June, July, Aug, Sept, Oct, Nov, Dec		
1997	Jan, Feb, Mar, Apr, May, June, July, Aug, Sept, Oct, Nov, Dec		
1998	Jan, Feb, Mar, Apr, June, July, Aug, Oct, Nov, Dec	Sept	
1999	Jan, Feb, Mar, Apr, May, June, July, Aug, Sept, Oct, Nov, Dec		

Operations and Maintenance

Greenville instituted a plan in 1994 to improve disinfection practices and initiate unidirectional flushing in the system. Greenville took several steps in order to get their coliform levels under control. They initiated a nitrification detection program and organized an on-going unidirectional flushing program. They also maintain their chloramine residual at a level greater than 2.0 mg/L and switch from chloramines to free chlorine for one month per year.

The unidirectional flushing program is year round and targets velocities of 2.5 ft/s. Water from galvanized pipe is especially difficult to clean up. Greenville has a long-term maintenance and capital improvement plan that targets the replacement of galvanized, cement, transite, and certain pre-stressed concrete cylinder pipe for replacement. Unlined cast iron pipe is replaced on an as-needed basis, dependent on its history and water quality. After a pipe replacement, coliform, total chlorine, HPC, and pH tests are performed. Pipe breaks usually occur due to pipe age, corrosion of material, inadequate depth of bury, and pressure transients. After a pipe break, hypochlorite is applied to the open trench area and to the interior of all pipes and fittings used in making the repair. The area of the break is flushed until discolored water is eliminated. Depending on the severity of the break, bacteriological samples are sometimes taken.

Tanks and reservoir are inspected bi-annually and cleaned if necessary. Within a year, the system no longer exceeded the total coliform concentration at more than 5% during any month and Greenville has been able to maintain control of coliforms at levels less than 5% since. The difficulty of eradicating regrowth once it has become established is demonstrated by the continued low level, less than 2.5%, of coliforms in the distribution system."

Greenville uses a hydraulic model of their system to provide an overall picture and gradients throughout the system under different flow conditions. They also have a formal customer complaint tracking procedure. The most common complaint is discolored water. When a complaint is called in, it is categorized and baseline information about the location, timing (intermittent or chronic), and particular customer observations are noted. Greenville responds on site to each complaint unless it is a "muddy water" complaint. Tests performed at the site are dependent on the nature of the complaint. In each case, chlorine residual, turbidity, pH and total coliform tests are done as a minimum. Other tests that may be run include color, iron, manganese, copper, zinc, microscopic particle analysis, heterotrophic plate count, hardness and alkalinity.

COLUMBUS, OHIO

Introduction

The City of Columbus primarily uses the Scioto River, a surface water source, to serve its 330,000 customers in 8 municipalities. Average day demand is 53 MGD. A small fraction (11%) of the water served comes from groundwater sources. Treated water storage capacity for the system is 65 million gallons.

Columbus uses conventional treatment of the surface water source, with chlorine as the primary and secondary disinfectants. The target disinfectant residual leaving the plant is 1.5 mg/L and the typical range of chlorine residual in the distribution system in 0.2-1.3 mg/L. Zinc orthophosphate is used at a dose of 1.5 mg/L as PO4 for corrosion control.

Columbus has a 3100 mile pipe network, with pipe sizes ranging from 66 to 1 ½ inches in diameter. The pipe make-up for the Columbus water distribution system is as follows:

- Lined ductile iron – 45%
- Unlined cast iron – 40%
- Galvanized steel – 8%
- Concrete - 7%

The oldest pipe is 121 years old and the average pipe age is reported as less than 10 years. Columbus was chosen for a case study because of its conventional treatment of a surface water source, large percentage of unlined cast iron pipe in the distribution system, and small number of coliform incidences.

Water Quality

Columbus tests for disinfectant residuals, HPCs, and coliforms daily and iron levels monthly. Turbidity is tested whenever samples are collected for lead and copper analysis. The typical age of water at the sample sites is 36 hours and the maximum age at one of these sites is 72 hours. Because of the small number of coliform incidences, it is impossible to say whether some sites are more prone to coliform positive samples than others.

The typical pH in the system is 7.8. Iron levels are very close to the minimum detectable, reportedly due to the high pH of the water. Parameters are almost never outside their normal range. In the rare case that a chlorine level is found to be 0, the City tests areas around the sample site. If it is not an isolated incident, the City looks for closed valves.

Columbus has had two coliform positive samples in the past nine years: one in January of 1991 and one in September of 1993. In both cases, a restaurant sink at the sampling site was implicated. After the first incidence, the sink was replaced. In the second case, the sink was very dirty. Additional samples were taken within the sample site buildings that were coliform positive and these repeat samples were negative. Sites upstream and downstream from the restaurant were also negative.

Since Columbus appeared to have determined the cause of their coliform positives, no treatment changes were made. Columbus did review their sampling site selection process and chose more appropriate sites for coliform sampling to minimize the possibility of sample contamination. These positive samples were clearly isolated incidents and not a pattern of chronic microbial presence in the distribution system.

Operations and Maintenance

Columbus has a vast rehabilitation and replacement program for their unlined cast iron pipe. They spend $5 million a year to replace the older pipe with bad break records, starting with the smaller diameter pipes. Their goal is to replace 8 miles of pipe a year. They also spend $1.5 million a year on cleaning and lining transmissions mains. In addition, they have a capital improvement plan that guides pipe replacement, tank maintenance, and other infrastructure work. Storage tanks are coated steel or iron and are drained and cleaned every two years.

Flushing is not done on a regular basis due to manpower constraints, but it is done to address specific problems. When a main is put into service, local flushing is performed and coliform sampling is done as needed. Main breaks usually occur due to corrosion of pipe material and pressure transients.

They also have a formal procedure for tracking customer water quality complaints, which are most often rusty water or taste and odor. Customer complaints are handled over the phone whenever possible, rather than at the site. If a customer complains of rusty water, it is usually because flushing or pipe repair is occurring in the area. If the complaint is taste and odor, it is usually due to algae blooms. If the complaint is strong chlorine odor, the City recommends that the customer put their water in a glass bottle in the refrigerator. Only when an unusual complaint comes in does the City make a trip out to the site for testing.

SCOTTSDALE, ARIZONA

Introduction

Scottsdale obtains 50% of its source water from surface water, 30% from groundwater, and purchases the remaining 20%. Some of the groundwater sources undergo air stripping to remove volatile organic contaminants. The population served is 200,000, which is all for one municipality. The average daily demand is 67 MGD, with 49.9 MG of reservoir storage.

For the surface water source, Scottsdale uses a conventional treatment process with chlorine as primary and secondary disinfectant. The groundwater is also chlorinated at the wells. Some of the newer wells have detention basins for chlorine contact. This has helped maintain the target residual. The target residual level leaving the plant is 1.3 and the typical range in the distribution system is 0.8-1.2. Corrosion control is normally not needed, but on occasion the pH is lowered with acid to 7.7.

Scottsdale's pipe materials consist of the following:

- Asbestos cement – 65%
- Plastic pipes – 30%
- Ductile iron – 2%
- Lined cast iron – 1%
-
- Steel cylinder – 1%

The majority of the pipes are 10-15 years old and the oldest pipe is 65 years old.

Scottsdale was chosen for a case study because of it has both groundwater and filtered surface water sources, lacks unlined cast iron pipe, and has been subject to a high number of coliform incidences in the past few years.

Water Quality

Since 1992, Scottsdale has had 30 incidents of coliform presence in the 0-2.5% range, and four incidents of coliform presence in the 2.5-5% range. A Total Coliform Rule violation, i.e., coliform in more than 5% of the monthly samples, has not occurred. Fecal coliform organisms were detected six times in 1998 and were detected once in 1999. Fecal coliform and E. coli records previous to 1998 are unavailable, although the person interviewed recalled a major fecal coliform problem in 1992 resulting in a boil order.

Review of the months of coliform incidence from 1992 through spring of 2000 reveals that the majority of coliform incidences and the higher levels of coliform incidence occur in the late summer and fall. Of the 34 incidents, 17 occurred in August, September, October and November. Thus, there seems to be a seasonal effect even though Scottsdale is in a relatively warm area.

Scottsdale tests its water for disinfectant residuals, HPCs, and coliforms weekly, and for pH and DBPs quarterly. They do not test for turbidity or iron. HPCs are only monitored at locations at 15 sites that have repeatedly had detectable levels. The 128 water quality sampling sites are spread out in the distribution system and are located in areas where asbestos cement and cast iron pipes are prevalent. The typical water age is 1 to 4 days at these sites, with a maximum age of 5 days. The locations that are more prone to coliform incidences are in dead end areas where customer complaints are highest. Typical water quality in the distribution system is presented in Table B.4.

If a parameter is outside the norm, the City responds by checking chlorine residuals and talking to operations staff to find out if there are closed valves or construction in the area. If the HPC values are above normal, the City flushes the area.

Table B.4
Distribution System Water Quality – Scottsdale, Arizona

Parameter	Minimum	Maximum	Average
pH	7.0	8.4	7.8
Chlorine Residual, mg/L	0.8	1.2	1
HPC, cells/mL	<1	>500	1-10

Sampling sites used to be at outdoor taps or hose bibs on residents' houses. Often, the sampler would note bird feces or other contaminants at the sample tap. Hence, many of the coliform incidents were attributed to contamination of the sample taps. In the past two years, the City changed all sample locations to dedicated enclosed sampling stations to remove external environmental contamination. The incidence of positive coliforms has decreased dramatically. The City also increased the average chlorine residual goal to 0.8 to 1.2 mg/L system wide from the prior target of 0.5 mg/L.

Operations and Maintenance

Scottsdale does not regularly flush its distribution system, mainly for community relations reasons. Since Arizona has such an arid climate, Scottsdale encourages its residents to conserve water whenever possible. It is feared that if residents witnessed the City flushing the system, there could be some public objection. Also, the City system is relatively new and the City doesn't see a need for regular flushing. Flushing does occur in dead end areas when it is needed, at a flow of 5 ft/sec for 6" pipe. Flushing also occurs whenever a main breaks and the repairs are complete. The water is re-chlorinated before the main is put back into service. Scottsdale reports many factors that contribute to main breaks, including pipe age, corrosion of pipe material, inadequate depth of bury of pipe, heavy loads, pressure transients, and heat impact due to high temperatures.

Scottsdale has a long-term maintenance plan for pipe replacement, tank maintenance, and other infrastructure work. About 1700 service lines, equal to 4900 feet, are replaced each year. Most (90%) of the service lines being replaced are PVC and 10% are galvanized iron. The service lines are replaced with copper piping. Water tanks are inspected monthly and cleaned as needed.

Scottsdale uses a hydraulic model for predicting future demands. They also have a formal procedure for tracking customer complaints, the most common of which are hardness, scaling, and spotting. They try to resolve these complaints over the phone but, if necessary, will send staff out to the site to take coliform or other water quality samples. They also print a brochure on hard water to help customers.

WATER DISTRICT NO. 1 OF JOHNSON COUNTY, KANSAS CITY, KANSAS

Introduction

Water District No. 1 of Johnson County serves 17 municipalities in the Kansas City, Kansas region. The average daily demand of the water system is 50 MGD, which serves a population of 357,645. The total volume of storage in the distribution system is 61.85 MG. Johnson County gets 97% of its water from surface water sources (the Kansas and Missouri Rivers) and 3% from groundwater.

Water District No. 1 use conventional treatment with lime softening. They use chlorine dioxide for primary disinfection and chloramines for secondary disinfection. Orthophosphate is added at a concentration of 0.25 mg/L and pH is adjusted to 9.5, prior to water exiting the plant. The target chlorine residual, both leaving the plant and in the distribution system, is 2.5-3.0 mg/L.

The distribution system is 2,200 miles in length and contains pipes ranging in size from 2 inches to 54 inches. Pipes in the District's distribution system are composed of the following materials:

- Cement lined ductile iron– 30%
- Cement lined cast iron– 30%
- Polyvinyl Chloride Pipe – 24%
- Prestressed Concrete Cylinder Pipe – 15%
- Unlined cast iron – 5%
- Asbestos Cement – 1%

The oldest pipe is 80 years old and the majority of the pipes are 30 years old.

Water District No. 1 was chosen for a case study because of its softened surface water source, small percentage of unlined cast iron pipe, and absence of coliform problems.

Water Quality

Water District No. 1 tests for disinfectant residuals, pH, HPCs, coliforms, and turbidity daily, and for iron approximately once a month. The typical age for sample sites in the system is 1 day and the maximum age is 2 days. They have had no positive coliforms in the past 9 years. Table B.5 presents typical water quality in the distribution system.

Table B.5
Distribution System Water Quality, Johnson County, Kansas

Parameter	Target	Minimum	Maximum	Average
PH	9.5	9.0	10.0	9.5
Chlorine Residual, mg/L	2.75	1.2	3.8	2.7
Turbidity	0.10	0.01	0.35	0.08

Operations and Maintenance

Water District No. 1 has a regular unidirectional flushing program using velocities of 5 to 10 ft/s. Some areas are flushed weekly in the summer. Flushing also occurs after a pipe is repaired or replaced. Water from unlined cast iron pipe takes longer to clean up.

Water District No. 1 also has a long-term maintenance plan. Tanks, which are made of coated steel, iron or concrete, are cleaned every 5 years and painted every 10 years. Pipes are replaced when they have a persistent water quality problem or when numerous breaks occur in a limited time. Breaks tend to occur due to corrosion of pipe material, pressure transients, and poor installation practices. Unlined cast iron pipe is budgeted to be replaced or relined over the next 10 years.

Water District No. 1 uses a hydraulic model of their system for design purposes. They also use a formal procedure to track customer complaints. The most common complaint is turbid water, which can be comprised of either air bubbles or rusty particles stirred up during repairs or disturbances to mains. When there is a complaint, representatives of Water District No. 1 visit the site to sample and try to resolve the complaint.

TOWN OF GILBERT, ARIZONA

Introduction

Gilbert obtains 52% of its water from surface water and 48% from groundwater. The population served is 100,000, all of whom are service within the Gilbert town limits. The average daily demand of the water system is 18 MGD. The total volume of storage is 26 MG.

The treatment process is conventional, with chlorine as both the primary and secondary disinfectants. The target residual leaving the plant is 1.2 to 1.5 mg/L and the residual range in the distribution system in 0.6 to 1.0 mg/L. Sodium hydroxide is used to adjust the pH to 7.8 to 8.0 for corrosion control.

The Gilbert distribution system pipe material breakdown is as follows:

- C900 Polyvinyl Chloride – 85%
- Asbestos Cement – 10%
- Prestressed Concrete Cylinder Pipe – 5%

There are no unlined cast iron pipes in the system. The majority of the pipes are 5 to 10 years old, and the oldest pipe is only 20 years old. The total length of piping in the distribution system is 650 miles and pipe diameters range from 4 inches to 48 inches.

Gilbert was chosen for a case study because it has a mixed groundwater and surface water source, no unlined cast iron pipe, and presence of coliform in the 0 to <2.5% range.

Water Quality

Gilbert tests for disinfectant residuals, turbidity, pH, and coliforms daily; iron levels, monthly; and HPCs, quarterly. The sampling sites are spread throughout the city and do not appear to play a role in coliform levels. Typical water quality is presented in Table B.6.

Coliform levels from 0 to <2.5% have been measured 24 times in 1998 and 1999. Fecal coliforms were present at these incidents. However, biochemical tests revealed that the coliforms were from an environmental source, not typical fecal coliforms, so no treatment changes were made in response to these incidents. Gilbert attributes the coliform levels to contamination with airborne coliforms, since no further contamination was found in the repeat samples.

Table B.6
Distribution System Water Quality – Gilbert, Arizona

Parameter	Minimum	Maximum	Average
pH	7.1	8.0	7.8
Chlorine Residual, mg/L	0.6	1.8	1.0
Turbidity, NTU	0.05	0.09	<0.09
Iron, mg/L	0.01	0.08	0.03
HPC, cells/mL			400

Operations and Maintenance

Gilbert does not regularly flush the distribution system; they do it on an as-needed basis. They do not have any pipeline replacement or rehabilitation plan. Pipelines are replaced after breaks, which are almost exclusively due to contractors hitting the lines. After a line is put back in service, it is re-chlorinated and tested for fecal coliforms.

There are four steel storage tanks and three concrete lined storage reservoirs in the system. They range in size from 1 million gallons to 16 million gallons. They are inspected on the outside yearly, but are checked thoroughly inside with divers every five years.

Gilbert has a hydraulic model of their system which they use as a water resources master plan. They also have a customer complaint tracking system; the most common complaints are odor and taste. Customers cite an earthy, musty odor and flavor. Depending on the complaint type, Gilbert may send a representative to the site to test chlorine residual and total coliform analysis, if necessary.

CITY OF KENNEWICK, WASHINGTON

Introduction

The City of Kennewick gets its water from two sources: 40% is from the Columbia River and 60% is from groundwater wells. The population served is 55,000, all within Kennewick City limits. The average daily demand of the system is 9 MGD and the total storage capacity is 21 MG, which includes 7 reservoirs. The ground water source is used throughout the year and is supplemented by the surface water from the Columbia River between April and November each year.

Chlorine is used as the only disinfectant for the groundwater wells. The groundwater sources also receive orthophosphate addition for corrosion control before the water is pumped to distribution. The surface water goes though a conventional treatment plant; however, the sedimentation basins are used only during high algae months. Ozone is used as the primary disinfection method at the filtration plant, with chlorine as the secondary disinfectant. The target disinfectant level leaving the plant is less than 1.0 mg/L and its range in the distribution system is 0.2 to 0.5 mg/L. No corrosion control is used for the surface water.

The pipe materials in the Kennewick distribution system are as follows:

- Cast iron – 10-15%
- Steel – 10-15%
- Asbestos cement – 40%
- Polyvinyl Chloride – 40%

The majority of pipes are 30 years old and the oldest pipe is 80 years old. The distribution system is composed of 380 miles of pipe, which range in diameter from 2 inches to 42 inches. There is one 500,000 steel storage tank and several concrete tanks which can store a total of 21 million gallons.

Kennewick was chosen for a case study because of its mixed source waters, small percentage of unlined cast iron pipe, and low-level coliform incidences.

Water Quality

Kennewick tests for disinfectant residuals, pH, and coliforms weekly. They test for iron levels, HPCs, and turbidity on an as-needed basis. Table B.7 presents typical water quality data.

Table B.7
Distribution System Water Quality – Kennewick, WA

Parameter	Minimum	Maximum	Average
PH	7.4	7.9	7.6
Chlorine Residual, mg/L	0.1	1.2	0.6
Turbidity, NTU	0.02	0.12	0.06
Iron, mg/L	<0.01	0.24	0.08
HPC, cells/mL	5	5	5

Coliform levels of 0-2.5% are detected once or twice per year, especially in the summer months. The sample sites where coliforms were detected are spread out in the distribution system. These incidents are usually attributed to the customer's tap, i.e. contamination of the faucet screen, or soft water conditioners that have not been serviced. A few have been attributed to sample contamination. Resampling, investigation of customers' premises, and interviewing customers and sampling personnel led to these conclusions.

Operations and Maintenance

Kennewick flushes their dead ends and mains twice a year. The length of the dead end, rather than the pipe material, tends to dictate how long the flushing water will take to clean up. Kennewick has a long term maintenance plan that includes annual tank inspection and cleaning, annual valve tuning, annual fire hydrant flushing, and capital prioritization and pipe replacement. Pipe material is one of the criteria used for rating pipes in the replacement program.

Pipes tend to break most often due to age and improper bedding during construction. Pipes are placed back in service following the procedure outlined in the Washington State Code.

Kennewick uses a hydraulic model of their system for capacity analysis and also to determine velocity to identify stagnant water. Kennewick also has a procedure for tracking customer complaints. The most common complaint is taste and odor from the summer algae. Ozone has not eliminated the algae problem during the summer months. The switch from the ground water source to a blend of surface and ground water in April makes the customers more apt to notice a difference in the taste of the water in the summer.

ORANGE WATER AND SEWER AUTHORITY, CARRBORO, NORTH CAROLINA

Introduction

Orange Water and Sewer Authority (OWASA) gets 100% of its water from surface water sources: University Lake Reservoir and Cane Creek Reservoir. Two municipalities are served, for a total population of 65,000 and an average daily demand of 8.7 MGD. The total volume of storage in the system is 8 MG.

Conventional treatment is used with chlorine as the primary disinfectant. The typical range of chlorine residual in the distribution system is from 0.2 to 1.5 mg/L. Poly- and orthophosphate are used for corrosion control at a concentration of 1.4 ppm.

The OWASA distribution system consists of 329 miles of pipe, which range in diameter from 2 inches to 24 inches. The pipe material breakdown is as follows:

- Unlined cast iron – 19%
- Asbestos cement – 55%
- Ductile iron – 19%
- PVC – 5%
- Other – 2%

The majority of pipes in the distribution system are around 40 years old, and the oldest is 80 years old. The storage tanks in the distribution system are made of coated steel or concrete.

Orange Water and Sewer Authority was chosen for a case study because of its filtered surface water, 19% unlined cast iron pipe, and low level coliform incidence.

Water Quality

Disinfectant residuals, pH, turbidity, HPCs, and coliforms are tested daily, Monday through Thursday. Iron levels are tested only when a customer requests them. The typical age at the sampling sites is 1 day, and the maximum water age is 3 days. Water quality is presented in Table B.8.

Table B.8
Distribution System Water Quality – OWASA, North Carolina

Parameter	Target	Minimum	Maximum	Average
PH	7.1-7.5	6.9	8.5	7.2
Iron, mg/L	<0.02	<0.02	<0.02	<0.02
Chlorine Residual, mg/L	>0.2	<0.20	2.0	1.0
HPC, cells/mL	0	0	500	1
Turbidity, NTU	<0.5	0.01	0.33	NA

OWASA uses the Colilert test to determine presence/absence of total coliform and *E coli*. Eight low-level coliform incidences (0-2.5%) have been recorded in the past nine years. One of the incidences was attributed to sampling error, but no cause was found for the other incidences. None of the coliforms were determined to be *E. coli*. OWASA isolated and re-chlorinated the area around the positive coliform samples. Coliform incidence was not confined to any particular area within the distribution system, nor was there a relation to a particular season or water temperature.

Operations and Maintenance

OWASA flushes portions of their distribution system monthly and when there are customer complaints. The portions that are flushed most frequently are problem areas, such as dead ends and areas with low chlorine residuals. The target velocities are around 3 to 5 ft/s.

OWASA also has a 15-year capital improvement plan: they will replace 1 mile of water lines in fiscal year 2000-2001, and increase to 3 miles a year by fiscal year 2004-2005. The plan does not specify any certain pipe material to have priority in replacement. Pipelines tend to break due to pipe age and corrosion of pipe material. When they break, OWASA uses AWWA guidelines to put the pipe back in service.

OWASA also has an annual storage tank maintenance contract: the interiors of the tanks are cleaned and inspected yearly and the interiors and exteriors are painted every 10 years. OWASA has a hydraulic model of their system which they use to find flushing residence time and for analyzing the system for capacity improvements. They also have a customer complaint tracking system. The most common complaint is odor.

BOSTON WATER AND SEWER COMMISSION

Introduction

The Boston Water and Sewer Commission provides drinking water to more than 500,000 people in Boston. The Boston Water and Sewer Commission purchases water wholesale from Massachusetts Water Resources Authority (MWRA) and delivers it to customers within the city boundaries.

The Boston Water and Sewer Commission receives unfiltered surface water that the Massachusetts Water Resources Authority has treated with free chlorine as a primary disinfectant and chloramines as the secondary disinfectant. The water is fluoridated and receives pH and alkalinity adjustment using sodium carbonate and carbon dioxide to render it less corrosive before MWRA provides it to wholesale customers such as Boston. The Boston Water and Sewer Commission does no further treatment of the water that it purchases.

Average daily demand for the Boston system is about 86 million gallons. This water is provided to customers though a 1,182 mile pipe network ranging in size from 48 inches (122 cm) to 4 inches (10 cm) in diameter and composed primarily of iron pipe. More than half (55%) of the distribution piping is unlined cast iron, more than a quarter (28%) is cement lined ductile iron, and the remainder is cement lined cast iron with a very small proportion (2%) of steel.

Overview of Regrowth Events

Starting in the spring of 1995, Boston Water and Sewer observed the presence of coliforms at distribution systems sampling points. Coliform numbers remained at levels of less than 5% until the summer of 1996 when the Boston Water and Sewer Commission detected total coliforms within the distribution system at numbers above the maximum contaminant level. Drinking water samples collected in June though August of 1996 from the City of Boston's water distribution system exceeded the Total Coliform maximum contaminant level of 5%. Coliforms were found throughout the distribution system, indicating a widespread problem, rather than an isolated incident. The majority of coliform type organisms were identified as *Klebsiella* species, *Enterobacter* species, or *Citrobacter*. *Klebsiella* species form a polysaccharide capsule that provides some protection from disinfection. *Enterobacter* and *Citrobacter* species are most often found the natural environment.

Boston Water and Sewer System Response

In September, 1996, Boston Water and Sewer Commission signed an Administrative Consent Order with Massachusetts Department of Environmental Protection. The Administrative Consent Order required Boston Water and Sewer Commission to develop and implement a plan to reduce coliform incidence. Boston Water and Sewer Commission hired a consultant to help them evaluate the causes of bacterial growth in the distribution system and form a plan to improve water quality. The evaluation revealed that most of the distribution sampling sites lacked a consistent chloramine residual above 0.20 mg/L. The warmer water temperatures of summer promoted the proliferation of bacteria while a change in corrosion control treatment by MWRA in June, 1996, may have destabilized the pipe surfaces, allowing the bacteria to release from the pipe walls. No correlation was found between pipe age and coliform incidence. Primary recommendations to the Commission were to:

1. Improve chlorine (chloramine) residual levels throughout the distribution system. A minimum level of 0.20 mg/L as chlorine was recommended to prevent microbial growth.
2. Institute a flushing program and conduct sampling before and after flushing to determine differences in water quality.
3. Refine analytical methods used to detect coliforms. In concert with MWRA, who performs the actual microbial analyses for Boston Water and Sewer, the Commission should use the membrane filter method of testing for coliforms so that fewer atypical species are found. All presumed coliforms should be identified to the species level.
4. Install dedicated sample stands rather than use schools, fire department buildings, or hospitals for coliform sample locations.

In 1997, Boston Water and Sewer Commission initiated a Biofilm Corrective Action Plan consisting of a comprehensive unidirectional water main flushing program, replacement of water mains and cleaning and cement lining of selected water mains. Between 1991 and 1997, Boston Water and Sewer Commission replaced or relined 106 miles of water mains and approximately 5,000 fire hydrants. In 1997, a total of 25 miles of pipe was replaced or relined. The goal is to have no pipe older than 100 years in the system by 2010. Currently, the oldest pipe in the distribution system is 140 years old.

The Massachusetts Water Resources Authority assisted Boston Water and Sewer Commission to solve the biofilm problem by modifying disinfection to increase chloramine residuals and implementing improved corrosion control. Boston Water and Sewer Commission monitors chloramine residuals and total coliform concentrations daily within the distribution system and monitors heterotrophic plate counts using the R2A method weekly. The Commission aims for a typical chloramine residual of 1.7 mg/L. Massachusetts Water Resources Authority also monitors pH at six locations within the city.

In 1998, within a year of implementation of the Biofilm Corrective Action Plan, incidence of coliform recovery in the distribution system dropped to less than 0.25 percent.

APPENDIX C
RECOMMENDED PROTOCOL FOR ANNULAR REACTOR STUDIES

The following protocol was recommended to each of the field study utilities for sampling and analyses of the coupons from the annular reactors. Utility personnel collected the coupon samples and all analyses were completed at Montana State University. Schematics of the two typical annular reactor designs used for this project are displayed in Figure C.1.

COUPON SAMPLING PROTOCOL

Required Equipment

- spray bottle containing disinfectant (70% EtOH is recommended but 5% bleach can be used but is extremely corrosive, and should be rinsed and wiped clean with a damp cloth after sample port has been closed)
- 100 ml sterile phosphate-buffered saline (PBS) or water (refrigerate until use; one per slide)
- disposable rubber gloves
- 30 ml wide mouth, screw-topped test tube (one per slide)
- sterile hook to retrieve the slide(s)
- sterile replacement slide(s)

Sampling Procedure

1. Turn off drive motor so inner drum is no longer spinning.

2. Fog the sampling port with disinfectant.

3. After approximately 20 seconds, quickly open the sampling port, taking care to leave it open only a minimal amount of time.

4. Quickly remove the desired coupon using a sterile metal hook or other type of device to slide the coupons out of the reactor. Drop the slide into the sterile wide mouth screw-topped test tube and fill with sterile PBS (or sterile water) prior to sealing for transport and analysis. If inner drum is not in proper position to remove desired coupon, rotate inner drum slowly by using the speed controller, noting the engraved numbers on the inner drum.

5. Record the coupon number and date sampled and replace with a sterile slide. Replace hole plug. Test tubes containing the coupons should be stored on ice. Coupon biofilm analysis should be conducted within 24 hours of sampling.

6. Repeat above procedures 1-5 for each slide to be removed.

7. Restart cylinder rotation.

COUPON ANALYSIS PROTOCOL

Required Equipment

- spray bottle containing disinfectant (70% EtOH is recommended but 5% bleach can be used but is extremely corrosive, and should be rinsed and wiped clean with a damp cloth after sample port has been closed)
- sterile cell scraper (one per slide)
- 100 ml sterile phosphate-buffered saline (PBS or water -- refrigerate until use; one per slide)
- sterile 200 ml beaker (one per slide)
- disposable rubber gloves
- sterile 5 ml pipet (one per slide)
- sterile tweezers (rinse in 705 ethanol and flame between each coupon handled)

Biofilm Removal

1. Put on rubber gloves, wash in disinfectant (70% ethanol; avoid flames!), allow to dry for 20 seconds.

2. Remove the coupon from the test tube using the sterile tweezers, and place upright in the 200 ml beaker.

3. Pour the contents of the test tube over the slide surface, being careful to contain all the liquid in the beaker.

4. Carefully lift coupon and scrape using sterile cell scraper into beaker. Rinse scraper in diluent. Repeat three times. (Repeat scraping and rinsing)

5. Using a 5 ml pipet, rinse scraped surface using sterile PBS. Repeat 4 times. Set coupon aside. Rinse cell scraper using the 5 ml pipet and sterile PBS into the 200 ml beak. Set scraper aside.

6. Decant the remainder of the 100 ml of sterile PBS into the test tube, emptying when full into the 200 ml beaker (total volume = 100 ml).

Scraping, Homogenizing, and Sampling Protocol

1. Homogenize the biofilm suspension fluid with a disinfected tissuemizer for one minute (either autoclave the tissuemizer probe or disinfect using 70% ethanol and sterile water rinse prior to homogenization of each sample).

2. Remove 1 ml of the fluid for a dilution series (plate 0.1 ml of the dilutions estimated to provide countable colonies on agar plates). Place each dilution of interest in triplicates (minimum).

3. Incubate for 2-5 days at 20°C.

4. Count plates containing 30-300 colonies.

5. Calculate surface cell densities from the plate count data. The surface area of the slides is determined by calculating: [(0.580 inches - 2 x.0424) x (5.90 inches) = 2.92 in^2 or 1884 mm^2 or 18.84 cm^2. Surface density calculations should be based on the biofilm scraped into a 100 ml volume, and sequential dilutions performed on the 100 ml solution.

Plating Protocol

Bacteria can be plate counted on selective agar or standard plate count agar (i.e. HPC or R2A; Standard Methods for the Evaluation of Water and Waste Water, 1989, 17th Edition, American Public Health Association, Port City Press, Baltimore, MD).

As a first approximation for plating, a monolayer surface coverage generally is on the order of 10^{10} to 10^{11} cells/m^2. A thin biofilm often contains 10^{11} to 10^{12} cells/m^2 and a thick biofilm on the order of 10^{12} to 10^{14} cells/m^2. When plating, several dilutions should be plated in triplicate to ensure valid plate count data is obtained. Often a dilution above and below the suspected bacterial dilution that will provide 30-300 cell colonies/plate, are plated. If no bacterial polymer/slime suspension is observed during scraping, the cell count will likely be less than 10^{10} cells/m^2.

Continuous Flow (Once Through)
Laboratory Annular Reactor System
Schematic

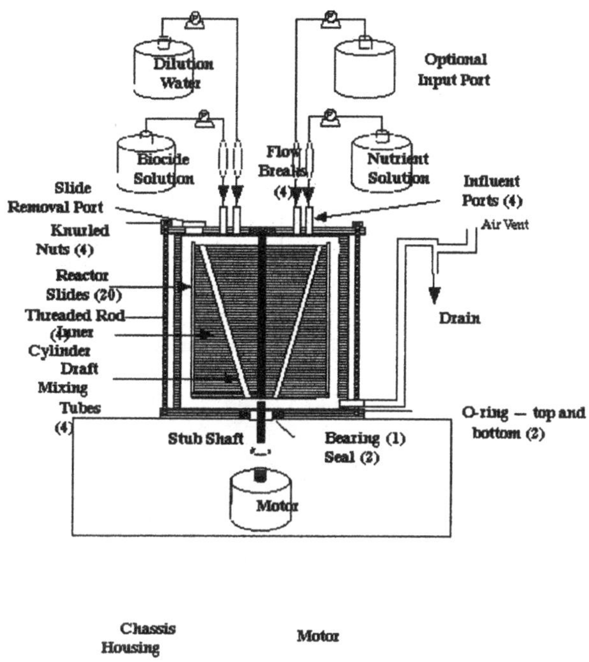

Figure C.1 Annular reactor schematic – bottom mount motor

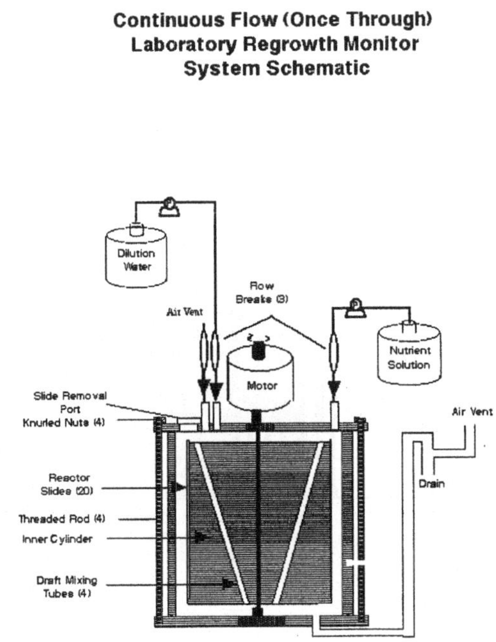

Figure C.2 Annular reactor schematic – top mount motor

APPENDIX D
PIPE EXCAVATION RESULTS AND PROTOCOL

DESCRIPTION OF PIPE EXCAVATIONS

In addition to the annular reactor field studies, sections of 12-in cast iron pipe were excavated from various locations in Cambridge Water Department's (CWD) and Greenville Water System's (GWS) distribution network during their annual spring/summer pipe replacement program. One cast iron pipe section was excavated from the CWD system and two cement-lined pipe sections were excavated from GWS. Maintenance crews from each utility performed the excavations and transport.

A general procedure was recommended to utility personnel for excavating distribution system pipe samples for bacterial scale analysis. During both the excavation and the shipping process, special care was taken not to strike or drop the pipes to preclude disturbing the interior corrosion scales. Each pipe was marked in the trench to indicate the orientation (both axial and top/bottom) of the pipe and the most common flow direction and the pipes were maintained full by capping them on each end during excavation. Pipes were sent to Dr. Anne Camper at Montana State University for analysis.

On July 13, 2000, two two-foot sections of 6-in diameter cement-lined pipe were cut from Greenville Water Department pipelines, which had been in service since 1972. All pipe sections were sent to Montana State University for analysis.

In July 2000, an 18-in section of 6-in diameter unlined cast iron pipe was excavated from the distribution system of the Cambridge Water Department. The pipe age was unknown, due to missing records. This pipe section was also sent to Montana State University for analysis. The results of the biological analyses completed on these pipe sections are summarized below, followed by a revised protocol for excavation and handling of pipe specimens and descriptions of the analytical procedure used to enumerate attached growth

RESULTS

The results for each pipe excavated in July are presented in Table D.1. The log attached biofilm levels from the cement lined pipe specimens from Greenville were higher than the levels measured from the annular reactor coupons (log HPC levels of 5.9 to 5.25 from the cement lined pipe specimens compared to average log HPC levels of 1.6 to 3.2 from the cement coupons). The cast iron pipe specimen from Cambridge exhibited biofilm HPC levels that were lower that what was observed on the iron coupons from the annular reactors at that utility.

Table D.1
Results of Pipe Excavation Analyses

	HPC (#/cm^2)	Log HPC	HPC (# gram dry mass)	Coliforms (#/cm^2)	Coliforms (#/gram dry mass)
Greenville					
Pipe #1 (cement lined)	8.2 x 104	4.91	5.1 x 107	1.6	1000
Pipe #2 (cement lined)	1.8 x 105	5.25	3.4 x 108	0	0
Cambridge					
Pipe top (cast iron	9.1 x 104	4.96	1.4 x 107	0	0
Pipe bottom (cast iron)	2.8 x 105	5.45	3.5 x 106	0	0

PIPE SAMPLE EXCAVATION PROCEDURE

The following is a general procedure for excavating distribution system pipe samples for bacterial scale analysis. This procedure can be used for unlined or cement-lined cast iron pipes. While no special method of pipe cutting is required, care should be taken to minimize disturbances to the pipe scale within the sample, during both the excavation and the shipping process.

Materials

Pipe Cutter or Saw
6"x4" Rubber Couplings w/hose clamps (Two per sample)
4-inch PVC plugs (Two per sample)
1 Flat-Head Screwdriver
1 Can Optic Orange Spray Paint
1 Roll Duct Tape
1 Black Magic Marker
Hot water (if accessible)

Sample Procedure

1. Follow utility standard operating procedures for isolating, bleeding and cutting the desired section.

2. Wherever possible, pipe samples should be taken from a straight section of a 6-inch line, at least 8 feet from any fittings or junctions. Take note of the condition of the pipe section.

3. Each sample should be between 9 and 15 inches long. The ideal length is 12 inches.

4. Throughout the process, special precautions should be taken to minimize disruption to the scale within the pipe sample. It is understood and accepted that the act of cutting the pipe may cause some disruption to the scale. Additionally, a backhoe may have to be used to pull out the pipe section if it is wedged in too tightly by the adjacent piping. The backhoe should pull out the pipe section as gently as possible and any additional disturbance should be kept to an absolute minimum.

5. Vertical pipe orientation must be indicated on the sample before cutting and maintained throughout the sampling and shipping process. If a backhoe is used, try to maintain the pipe orientation as much as possible. A spray-painted "T" on the top of the pipe is appropriate for labeling. If the spray-paint becomes illegible during excavation, rewrite the information on duct tape and attach to pipe.

6. The primary flow direction must be indicated on the pipe sample. A spray-painted arrow usually works well. If the arrow becomes illegible during excavation, redraw the arrow on duct tape and attach to pipe.

7. After cutting, the sample should be capped and filled with tap water from the same line if possible, or nearby line or hydrant. The caps are constructed of 6-inch x 4-inch rubber reducer couplings, with 4-inch PVC plugs in the small end. The large end of the coupling will stretch over the open end of the pipe (although the rubber may need to be submersed in hot water for several minutes beforehand in order for it to stretch sufficiently). If hot water is unavailable, the rubber can still stretch over the pipes if half of the coupling is held in place and a screwdriver is used to pry the other half over the edges of the pipe. This effort may require two people.

Note: A vise is recommended for holding the sample section while the end caps are stretched on. If no vise is available, the pipe may be placed up against a stationary support while the caps are put on.

Packing and Shipping

1. Pipe samples must be packaged carefully in a crate or cooler so that the tops of the samples remain facing up.

2. The samples should be packed so that they are prevented from rolling or shifting in transit.

3. The shipping container should have a "This Side Up" label on the top face.

Analytical Procedure

Pipes were sent to Dr. Anne Camper of Montana State University for analysis. The analytical procedure used is as follows:

1. A 25 cm^2 template was placed over the pipe.
2. The material within this template area was scraped and placed in 40 ml of sterile water.
3. The material and sterile water mixture was homogenized at maximum speed for 60 seconds to disperse the deposits.
4. The material was diluted as necessary and then spread plate in triplicate on R2A agar and incubated for 7 days at room temperature.
5. Triplicate spread plates for coliforms on mT7 incubated at 35^0 C for 24 hours were also done.
6. The remaining sample was used to determine the dry mass in one ml of inoculum so that results could be expressed as both # /cm^2 and #/gram dry mass.

REFERENCES

Abernathy, Calvin G. and Anne K. Camper. 2001. *Interactions Between Pipe Materials, Organics, Corrosion Inhibitors, and Disinfectants on Distribution Biofilms. Volume 1: Bench and Pilot-scale Studies*. National Water Research Institute. Fountain Valley, CA.

Abernathy, Calvin Glenn. 1998. *The Effect of Corrosion Control Treatments and Biofilm Disinfection on Unlined Ferrous Pipes*. Ph.D. Thesis, Montana State University, Bozeman, MT.

Allen, M.J. and E.E. Geldreich. 1977. "Distribution Line Sediments and Bacterial Regrowth." In: *Proc. Water Quality Tech. Conf., Kansas City, MO*. American Water Works Association, Denver, CO.

Allen, Martin J., Raymond H. Taylor, Edwin E. Geldreich. 1980. "The Occurrence of Microorganisms in Water Main Encrustations." *Jour. AWWA*. (11): 614 – 625.

AWWA and AwwaRF (American Water Works Association and Awwa Research Foundation). 1992. *Water Industry Data Base: Utility Profiles*. Denver, Colo.: AWWA.

Block, J.C., K. Haudidier, J.L. Paquin, J. Miazga and Y. Levi. 1993. Biofilm Accumulation in Drinking Water Distribution Systems. Biofouling 6:333-343.

Buls, J. 2000. Making Humic Substances More Biodegradable. M.S. Thesis. Montana State University, Bozeman, MT

Camper, A., M. Burr, B. Ellis, P. Butterfield, and C. Abernathy. 1999. Development and Structure of Drinking Water Biofilms and Techniques for Their Study. *Journal of Applied Microbiology Symposium Supplement*, 85: 1S-12S.

Camper, A.K., W.L. Jones, J.T. Hayes. 1996. Effect of Growth Conditions and Substratum Composition on the Persistence of Coliforms in Mixed-Population Biofilms. *Applied and Environmental Microbiology*, 4014-8.

Camper, Anne K. 1994. "Coliform Regrowth and Biofilm Accumulation in Drinking Water Systems: A Review." *In Biofouling and Biocorrosion in Industrial Water Systems*. Edited by G.G. Geesey, Z. Lewandowski, and H.C. Flemming. Boca Raton, FL: Lewis Publishers.

Camper, Anne K. 1996. *Factors Limiting Microbial Growth in Distribution Systems: Laboratory and Pilot-Scale Experiments*. American Water Works Association Research Foundation. Denver, CO.

Characklis, W.G. and K.C. Marshall. 1990. *Biofilms*. New York: John Wiley & Sons, Inc.

Cohn, Perry D., Michael Cox, Paul S. Berger. 1999. Health and Aesthetic Aspects of Water Quality. In: Letterman, Raymond D. (ed), *Water Quality and Treatment*. McGraw-Hill, New York. Chapter 2: p 2.1-2.14.

Costerton, J. William, K.J. Cheng, Gill G. Geesey, Timothy I. Ladd, J. Curtis Nickel, Mrinal Dasgupta, Thomas J. Marrie. 1987. Bacterial Biofilms in Nature and Disease. *Annual Review of Microbiology*, 41: 435-464.

Cragnolino, G. and O.H. Tuovinen. 1984. "The Role of Sulphate-Reducing and Sulphur Oxidizing Bacteria in the Localized Corrosion of Iron-base Alloys. A Review." *Int. Biodet.*, 20, 9.

Davis, James A. 1982. "Adsorption of Natural Dissolve Organic Matter at he Oxide/Water Interface." *Geochimica et Cosmochimica Acta*. Vol. 46, pp 2381 - 2393.

Delenoue A, Holt, D, Woodward J.C, McMath S.M., Smith S.E. 1997. "Effect of Pipe Materials on Biofilm Growth and Deposit Formation in Water Distribution Systems." In: *Proceedings AWWA Water Quality Technology Conference*. Denver CO: AWWA.

Donlan, R.M. and W.O. Pipes. 1988. "Selected Drinking Water Characteristics and Attached Microbial Population Density." *Jour. AWWA*. 80(11):70-76.

Donlan, R.M., W.O. Pipes, and T.L. Yohe. 1994. "Biofilm Formation on Cast iron Substrata in Water Distribution Systems." *Water Res.*, 28(6):1497-1503.

dos Santos Afonso, Maria and Werner Stumm. 1992. "Reductive Dissolution if Iron (III) Hydroxides by Hydrogen Sulfide". *Langmuir, (* 8):1671-1675.

Fletcher, M. and K.C. Marshall. 1982. "Are Solid Surfaces of Ecological Significance to Aquatic Bacteria?" In: K.C. Marshall (ed.). *Advances in Microbial Ecology, Vol 6.* Plenum Press. New York, NY. 199-236.

Fransolet, G., G. Villers, and W.J. Masschelein. 1985. "Influence of Temperature on Bacterial Development in Waters." *Ozone Sci.*, 7(3): 205-227.

Geesey, G.G. 1991. "What is Biocorrosion?" *Biofouling and Biocorrosion in Industrial Water Systems*. G.G. Geesey and H.C. Flemming, Editors. Springer-Verlag, Berlin.

Geesey, G.G., L. Jang, J.G. Jolley, M.R. Hankins, T. Iwaoka, P.R. Griffiths.1989. Binding of Metal Ions by Extracellular Polymers of Biofilm Bacteria. *Water Science and Technology*, 20(11/12): 161-165.

Geesey, Gill G., Iwona Beech, Philip J. Bremer, Barbara J. Webster, D. Bret Wells. 2000. Biocorrosion. In: Bryers, James D. (ed), *Biofilms II: Process Analysis and Applications.* Wiley-Liss, Inc. 281-325.

Geesey, Gill. 1991. What is Biocorrosion? In: Flemming, H.C. and G.G. Geesey (eds). *Biofouling and Biocorrosion in Industrial Water Systems.* Springer-Verlag, Berlin. 155-164.

Geldreich. E. E. 1986. "Edited Transcript of the Workshop on Coliform Noncompliance Nightmares: Scenarios and Action Plans." *Proc. 1986 Water Quality Technology Conference.* American Water Works Association. Denver. CO.

Griebe, Thomas, Ching-I Chen, Rohini Srinivasan, Philip Stewart. 1994. Analysis of Biofilm Disinfection by Monochloramine and Free Chlorine. In: Geesey, G.G., Z. Lewandowski, H. C. Flemming, (eds). *Biofouling and Biocorrosion in Industrial Water Systems.* Lewis Publishers, p. 151-161.

Fransolet, G., G. Villers, and W.J. Masschelein. 1985. "Influence of Temperature on Bacterial Development in Waters." *Ozone Sci.*, 7(3):205-227.

Gu, Baohua; Jurgen Schmitt; Zhihong Chen; Liyuan Liang; and John F. McCarthy. 1994. Adsorption and Desorption of Natural Organic Matter on Iron Oxide: Mechanisms and Models. *Environmental Science and Technology*, 28:38-46.

Gu, Baohua; Jurgen Schmitt; Zhihong Chen; Liyuan Liang; and John F. McCarthy. 1995. Adsorption and Desorption of Different Organic Matter Fractions on Iron Oxide. *Geochimica et Cosmochimica Acta*, 59: 219-229.

Gu, Baohua; Tonia L. Mehlhorn; Liyuan Liang; and John F. McCarthy. 1996. Competitive Adsorption, Displacement, and Transport of Organic Matter on Iron Oxide: I. Competitive Adsorption. *Geochimica et Cosmochimica Acta*, 60:1943-1950.

Hanson, H. G., L.M. Mueller, S.S. Hasted, and D. R. Goff. 1987. *Deterioration of Water Quality in Distribution Systems.* American Water Works Association, Denver, CO.

Haudidier, K., J.L. Paquin, T. Francais, Pl Hautermann, G. Grapin F. Colin, M.J. Jourdain, J.C. Block J. Cheron, O. Pascal, Y. Leui and J. Miazga. 1988. Biofilm growth in a drinking water network: a preliminary industrial pilot plant experiment. *Wat. Sci. Technol.* 20:109-115.

Herson, D.S., D.R. Marshall, K.H. Baker, H.T. Victoreen. 1991. Association of Microorganisms with Surfaces in Distribution Systems. *Jour. AWWA.* 83:103-106.

Huck, P. 1990. Measurement of Biodegradable Organic Matter and Bacterial Growth Potential in Drinking Water. *Jour. AWWA*, 82:78.

Kerr, C.J., K.S. Osborn, G.D. Robson, and P.S. Handley. 1999. The Relationship Between Pipe Material and Biofilm Formation in a Laboratory Model System. *Journal of Applied Microbiology Symposium Supplement.* Society for Applied Microbiology. 85: 29S-38S.

Kirmeyer, G.J., J. Clement, A.M. Sandvig, G. Pierson, V. Snoeyink, A. Camper. 2001. *Distribution System Water Quality Changes following Corrosion Control Strategies.* Awwa Research Foundation. Denver, CO.

LeChevallier, Mark. 1989. Treatment to Meet the Microbiological MCL in the Face of a Coliform Regrowth Problem. *Proc. AWWA Water Quality Tech. Conf. Philadelphia, PA.* American Water Works Association, Denver, CO.

LeChevallier, M.W. 1990. Coliform Regrowth in Drinking Water: A Review. *Jour. AWWA* . 76-86.

LeChevallier M.W. 1997. Disinfection of Biofilms in Drinking Water Systems. In: *Proceedings AWWA Water Quality Technology Conference.* AWWA. Denver, CO.

LeChevallier, Mark; T.S. Babcock, and R.G. Lee. 1987. Examination and Characterization of Distribution System Biofilms. *Appl. Environ. Microbiol.* 53(12):2714-2724.

LeChevallier M.W., C.D. Cawthon, and R.G. Lee. 1988a. Factors Promoting Survival of Bacteria in Chlorinated Water Supplies. *Appl. Environ. Microbiol.*, 54(3):649-654.

LeChevallier, Mark W., Cheryl D. Cawthon, and Ramon G. Lee. 1988b. Inactivation of Biofilm Bacteria. *Applied and Environmental Microbiology*, 54(10): 2492-2499.

LeChevallier, Mark, Cheryl D. Lowry, Ramon G. Lee. 1990a. Disinfecting Biofilms in a Model Distribution System. *Journal AWWA*, 82(7): 87-99.

LeChevallier, M.W., W.H. Schulz, R.G. Lee. 1990b. Bacterial Nutrients in Drinking Water. In: *Assessing and Controlling Bacterial Regrowth in Distribution Systems.* American Water Works Association and Awwa Research Foundation. Denver, CO.

LeChevallier, M.W., W.H. Schulz, and R.G. Lee. 1991a. "Bacterial Nutrients in Drinking Water." *Applied Environmental Microbiology.* 57: 857-862.

LeChevallier, M.W., R.G. Lee, J.S. Young. 1991b. *Relationship Between Corrosion and Disinfection of Biofilm Bacteria.* American Water Works Service Company, Inc.

LeChevallier, Mark W., Cheryl D. Lowry, Ramon G. Lee, Donald L. Gibbon. 1993. Examining the Relationship Between Iron Corrosion and the Disinfection of Biofilm Bacteria. *Journal AWWA,* 85(7): 111-123.

LeChevallier, Mark W., Nancy J. Welch, Darrell B. Smith. 1996. Full-Scale Studies of Factors Related to Coliform Regrowth in Drinking Water. *Applied and Environmental Microbiology.* 62(7):2201-2211.

LeChevallier, MW., C.D. Norton, A. Camper, P. Morin, B. Ellis, W Jones, A Rompre, M. Prevost, J Coallier, P. Servais, D. Holt, A. Deanoue, and J. Colbourne. 1998. *Microbial Impact of Biological Filtration.* Awwa Research Foundation. Denver, CO.

Little, B.J., Wagner, P.A., Gerchakov, S.M., Walch, M., and Mitchell, R. 1986. Probing Microbiologically Induced Corrosion. *Corrosion*, 42, 689

Lowther, E.D. and R.H. Moser. 1984. Detecting and Eliminating Coliform Regrowth. *Proceedings of the AWWA Water Quality Technology Conference, Denver, CO.* American Water Works Association. Denver, CO.

Lui, W, H. Wu, Z. Want, S.L. Ong, J.Y. Hu, W.J. Ng. 2002. Investigation of Assimilable Organic Carbon (AOC) and Bacterial Regrowth in Drinking Water Distribution System. *Water Research*, 36:891-898.

Martin, R.S., W.H. Gates, R.S. Tobin, D. Grantham, P. Worlfe, P. Forestall, R. Sumarah. 1982. Factors Affecting Coliform Bacteria Growth in Distribution Systems. *Jour. AWWA.* 74:34-37.

McFeters, G.A. and A.K. Camper. 1988. Microbiological Analysis and Testing. *Bacterial Regrowth in Distribution Systems.* AWWA Research Foundation and American Water Works Association. Denver, CO.

McMath, S.M., A. Delanoue, D.M. Holt. 1997. 'Clumps' Shed from Pipewalls in Distribution Systems. In: *Proceedings AWWA Water Quality Technology Conference.* American Water Works Association. Denver, CO.

Neden, Douglas G., Robert J. Jones, Judy R. Smith, Gregory J. Kirmeyer, Glenn W. Foust. 1992. Comparing Chlorination and Chloramination for Controlling Bacterial Regrowth. *Journal AWWA* (84)7: 80-88.

Niquette, Patrick, Pierre Servais, and Raoul Savoir. 2000. Impacts of Pipe Materials on Densities of Fixed Bacterial Biomass in a Drinking Water Distribution System. *Water Resources* 34(6): 1952-1956.

Olson, B.H. 1982. *Assessment and Implications of Bacterial Regrowth in Water Distribution Systems.* EPA-600/52-82-072. U.S. Environmental Protection Agency. Washington, D.C.

Opheim, d., J. Grochowski, and D. Smith. 1988. Isolation of Coliforms from Water Main Tubercles, N-6. *Abst. Annual Meet. Amer. Soc. Microbiol.*, p. 245.

Qi, Liwei. 1999. *The Surface Interactions Among Iron Oxides, Humic Substances, and Biofilms.* Master's Thesis, Montana State University, Bozeman, MT.

Rice, E.W., P.V. Scarpino, D.J. Reasoner, G.S. Logsdon, D.K. Wild. 1991. Correlation of Coliform Growth Response with Other Water Quality Parameters. *Jour. AWWA,* 83:98-102.

Ridgway, H.F. and B.H. Olson. 1981. Scanning Electron Microscope Evidence for Bacterial Colonization of a Drinking Water Distribution System. *Applied Environmental Microbiology.* 41:274-287.

Rosenzweig. W.D. 1987. *Influence of Phosphate Corrosion Control Compounds on Bacterial Regrowth.* EPA CR-B11613-01-0. U.S. Environmental Protection Agency. Cincinnati, OH.

Sanderson, Sara S., Philip S. Stewart. 1997. Evidence of Bacterial Adaptation to Monochloramine in *Pseudomonas aeruginosa* Biofilms and Evaluation of Biocide Action Model. *Biotechnology and Bioengineering*, 56(2): 201-209.

Servais, P., P. Laurent, and G. Randon. 1993. Impact of Biodegradable Dissolved Organic Carbon (BDOC) on Bacterial Dynamics in Distribution Systems. In: *Proc. Water Quality Technology Conference.* Denver, Colo.: American Water Works Association.

Smith, D.B., A. Hess, S. Hubbs. 1990. Survey of Distribution System Coliform Occurrence in the United States. *Proc. AWWA Water Quality Technology Conference.* American Water Works Association. Denver, CO.

Smith, D.B., A.F. Hess, and D. Opheim. 1989. Control of Distribution System Coliform Regrowth. In *Proc. Water Quality Technology Conference*, Denver, Colo.: American Water Works Association.

Srinivasan, Rohini, Philip S. Stewart, Thomas Grieve, Ching-I. Chen, Xiaoming Xu. 1995. Biofilm Parameters Influencing Biocide Efficacy. *Biotechnology and Bioengineering*, 46: 553-560.

Stewart, P. S., L. Grab, J.A. Diemer. 1998. Analysis of Biocide Transport Limitation in an Artificial Biofilm System. *Journal of Applied Microbiology*, 85:495-500.

Tipping, E. 1981. Adsorption by Goethite of Humic Substances from Three Different Lakes. *Chemical Geology*. 33, 81-89.

Tuovinen, O.H. and Jason C. Hsu. 1982. Aerobic and Anaerobic Microorganisms in Tubercles of the Columbus, Ohio Water Distribution System. *Applied and Environmental Microbiology*. Vol. 44, No. 3. pp. 761-764.

U.S. Environmental Protection Agency. 1992. *Seminar Publication: Control of Biofilm Growth in Drinking Water Distribution Systems.* EPA/625/R-92/001. Office of Research and Development. Washington DC.

van der Kooij, D. 1992. Assimilable Organic Carbon as an Indicator of Bacterial Regrowth. *Jour. AWWA*, 84:57.

van der Kooij, D. and W.A. Oorhuizen. 1997. Biofilm Formation in a Distribution System Supplied with Unchlorinated Slow Sand Filtrate. In*: Proceedings of the AWWA Water Quality Technology Conference.* American Water Works Association. Denver, CO.

van der Wende, E. W.G. Characklis, and J. Grochowski. 1988. Bacterial Growth in Water Distribution Systems. *Water Science and Technology*, 20(11/12): 521-425.

van der Wende, E. W.G. Characklis, and D.B. Smith. 1989. Biofilms and drinking water quality. Wat. Res. 23:1313-1322.

van der Wende E., Characklis W.G. 1990. Biofilms in Potable Water Distribution Systems. In: McFeters, Gordon A., editor. *Drinking Water Microbiology.* New York, New York. Springer-Verlag. P. 249 – 268.

Victoreen H.T. 1977. Water Quality Deterioration in Pipelines. *Proc. Water Quality Tech. Conf., Kansas City, MO.* American Water Works Association. Denver, CO.

Victoreen H.T. 1984. The Role of Rust in Coliform Regrowth. *Proc. AWWA Water Quality Tech Conf.*, Denver, CO.

Victoreen, H.T. 1980. The Stimulation of Coliform Growth by Hard and Soft Water Main Deposits. *Proc. AWWA Water Quality Tech. Conf. Miami Beach, FL.* American Water Works Association. Denver, CO.

Videla, H.A. 1996. *Manual of Biocorrosion.* Lewis Publishers, Boca Raton, FL.

Volk, Christian, Esther Dundore, John Schiermann, Mark LeChevallier. 2000. Practical Evaluation of Iron Corrosion Control in a Drinking Water Distribution System. *Water Resources* 34(6): 1967-1974.

Xu, Xiaoming, Philip S. Stewart, Xiao Chen. 1996. Transport Limitation of Chlorine Disinfection of *Pseudomonas aeruginosa* Entrapped in Alginate Beads. *Biotechnology and Bioengineering*, 49: 93-100.

ABBREVIATIONS

ANOVA	analysis of variance
AOC	assimilable organic carbon
AWWA	American Water Works Association
AwwaRF	Awwa Research Foundation
BAC	biologically activated carbon
BAF	biologically activated filtration
BDOC	biodegradable dissolved organic carbon
BOM	biodegradable organic matter
^oC	degrees celsius
cfu	colony forming units
cfu/ml	colony forming units/milliliter
cm^2	centimeter squared
C:N:P	carbon:nitrogen:phosphorus
CPVC	chlorinated polyvinyl chloride
CWD	Cambridge Water Department
CWW	Cincinnati Water Works
DAF	dissolved air flotation
DO	dissolved oxygen
DOC	dissolved organic carbon
^0F	degrees Fahrenheit
Ft	feet
ft/sec	feet per second
GAC	granular activated cargon
g	gram
GWS	Greenville Water System
HPC	heterotrophic plate count
L	liter
MDPE	medium density polyethylene
mg	milligram
MG	million gallons
MGD	million gallons per day
mg/L	milligram per liter
MIC	microbiologically influenced corrosion
mL	milliliter
ml/min	milliliter per minute

mpy	mils per year
MWRA	Massachusetts Water Resources Authority
N	normal
N/A	not applicable
NMRD	Modified Robbins Device
NTU	nephelometric turbidity units
$\#/cm^2$	number per centimeter squared
#/g	number per gram
OWASA	Orange Water and Sewer Authority
PBS	phosphate buffer solution
ppb	parts per billion
ppm	parts per million
PE	polyethylene
PVC	polyvinyl chloride
SRB	sulfate reducing bacteria
TDC	total direct counts
TDS	total dissolved solids
TOC	total organic carbon
uPVC	unplasticized polyvinyl chloride
µm	micrometer
µg/L	microgram per liter
UV	ultraviolet

IWA Publishing's authorised EU representative for General Product Safety Regulations is Diane D'Arras, 15 rue Duret, 75116 Paris, France, e-mail: safety@iwap.co.uk.

Printed and bound by CPI Group (UK) Ltd, Croydon, CR0 4YY
17/09/2026
02218813-0019